CLIMATE CHANGE, SUSTAINABILITY AND CONSUMER

ABOUT THE EDITORS

Prof. Suresh Misra is Chair Professor (Consumer Affairs) and Coordinator, Centre for Consumer Studies at IIPA, New Delhi. He holds his Masters in Political Science from Jawaharlal Nehru University, New Delhi, Masters in Human Resource Management from Pondicherry University and D.Phil. from Allahabad University. He had his advance training in Corporate Management at AOTS, Japan, TQM at U.K. Civil Service College and in Training Techniques and Methodology from the Thames Valley University, UK. Dr. Misra has authored/edited 20 books and contributed several papers in journals of repute. His major areas of interest include Consumer Welfare, TQM, Public Sector Reforms, Development Administration and Local Governance.

Dr. Mamta Pathania holds her Masters, M. Phil. and Ph.D. in Public Administration from Himachal Pradesh University, Shimla. Currently she is working as Assistant Professor, Public Administration and Co-Project Director of National Consumer Helpline set up by Department of Consumer Affairs, GoI at Centre for Consumer Studies, Indian Institute of Public Administration (IIPA), New Delhi. Dr. Pathania has authored/edited 12 books, contributed several papers in journals and books of repute and participated in a number of seminars and conferences.

She is also a member of the Script Committee of the Department of Consumer Affairs and a Member of the National COPOLCO Mirror Committee of Bureau of Indian Standards (BIS). Her major areas of interest include Consumer Studies, Good Governance, Public Sector Reforms, Local Governance, Rural Development, HRD and Organization Behaviour.

Climate Change, Sustainability and Consumer

Towards a Better Future

Editors

Suresh Misra
Mamta Pathania

CONCEPT PUBLISHING COMPANY PVT. LTD.
NEW DELHI-110059

ISBN-13: 978-93-88937-58-0

First Published 2021

Published and Printed by

Concept Publishing Company Pvt. Ltd.
A/15-16, Commercial Block, Mohan Garden,
New Delhi-110059 (INDIA)
T: 91 11 25351460, 91 11 25351794
E: publishing@conceptpub.com; *W*: www.conceptpub.com

Cataloging in Publication Data--Courtesy: D.K. Agencies (P) Ltd. <docinfo@dkagencies.com>

National seminar on "Sustainability Consumption, Climate Change and Consumers—Towards a Better Future" (2010 : New Delhi, India)

Climate change, sustainability and consumer : towards a better future / editors, Suresh Misra, Mamta Pathania.

pages cm

"The book is an outcome of a National Seminar on "Sustainability Consumption, Climate Change and Consumers—Towards a Better Future", organized by the Centre for Consumer Studies, Indian Institute of Public Administration, New Delhi"—About the book.

Includes bibliographical references and index.

ISBN 9789388937580

1. Climatic changes—India—Congresses. 2. Sustainability—India—Congresses. 3. Consumers—India—Attitudes—Congresses. I. Misra, Suresh, editor. II. Pathania, Mamta, 1978- editor. III. Indian Institute of Public Administration. Centre for Consumer Studies, organizer. IV. Title.

LCC QC903.2.I4N38 2010 | DDC 363.738740954 23

भारतीय लोक प्रशासन संस्थान
Indian Institute of Public Administration
Indraprastha Estate, Fling Road, New Delhi-110002
www.iipa.org.in

Surendra Nath Tripathi, IAS (Retd.)
Director
(Former Secretary, M/o Parliamentary Affairs, GoI)

Foreword

Climate change has been widely recognised as one of the most significant challenges facing the global economy. In recent years, issues such as climate change, environmental degradation and sustainable consumption have almost become inescapable. Climate change brings warmer temperatures, increased frequency and severity of extreme weather events, and decreased availability of natural resources. Today globally people are concerned about climate change and its effects and many of them want to act in ways consistent with the necessary reduction in emissions Inspite of global awareness about the problem, there is a lack of effective action.

The concept of sustainable consumption is easily undestandable in terms of the statement of Mahatma Gandhi that *"The world has enough for everyone's need, but not enough for everyone's greed."* Sustainable consumption can be seen both as a challenge and an opportunity. Ever increasing and irresponsible consumption is putting a major strain on the environment, by using up resources many of which are finite and not easily renewable; by causing pollution and global warming induced by climate change; by adversely affecting the ecosystem; and by undermining sustainable lifestyles. Promoting sustainable consumption and production are important aspects of sustainable development, which depends on achieving long-term economic growth that is consistent with environmental and social needs. A key issue in such a scenario will be not just reducing the volume or quantity of goods and services consumed, but also the extent to which improvement in environmental quality and in human lives can be achieved through the substitution of more efficient and less polluting goods and services.

Although environmental awareness is increasing, most people find it difficult to relate their personal consumption habits to large scale problems such as climate change. Ant this a major cause of concern. We, as individuals, as individuals, citizens fail to link our individual as well as group action with such macro phenomenon as climate change and think that it is something which is impacted only by actions of say large MNCs or of government.

However, it is individual action, individual consumption decisions which ultimately add up and affect the region, the country and ultimately the world Hence each individual is important. In this respect it is the consumer and citizen who is the driver of all activity. It is the consumer or citizen who is king. Fundamental changes are needed in how natural resources are extracted, the way products are produced, distributed, used and disposed of. Action is needed at all levels of society- from individuals, industry, governments at central, state and local levels, and civil society organisations; in order to reverse current trends and to ensure that we live within ecological limits. To reverse the current unsustainable trends, people need to be well informed, empowered and to realize and feet that their actions make a difference. People need clear and better information about which changes it their spending and consuming habits will have the biggest impact They need to know which are the best-performing products, how to make the most efficient use of them and the cleanest ways to dispose of the resultant wastes.

I congratulate Prof. Suresh Misra and Dr. Mamta Pathania of Centre for Consumer Studies, I Indian Institute of Public Administration, New Delhi for bringing out this edited volume on a contemporary theme that concerns all of us. I am sure the readers will find it useful and enriching.

New Delhi
October 6, 2020

Surendra Nath Tripathi

Preface

The planet is under stress—be it access to water, deforestation, pollution, or greenhouse gas emissions. The world is facing a critical challenge. Increasing greenhouse gas emissions are raising the earth's average temperature. As a result, dramatic climate change is forecast and global scientific opinion predicts enormous developmental, economic, social and environmental stresses on our planet. Leading climate change experts have put forward a series of practical solutions to tackle climate change. These solutions include the technical standards published by the world's three leading international standardization organizations: the International Electrotechnical Commission (IEC), the International Organization for Standardization (ISO) and the International Telecommunication Union (ITU). In its groundbreaking report published in 2007, the Intergovernmental Panel on Climate Change (IPCC) cited technical standards, like those published by the IEC, ISO and ITU, as a means of mitigating climate change now, while offering the potential to reduce its effects in the future as new technologies are developed and mature. The three global organizations are coordinating their work to ensure that government, business and society are provided with the necessary tools to help combat global climate change and to support the reduction in greenhouse gas emissions by increasing energy efficiency, while facilitating sustainable development.

The standards produced by the three organizations reach across all sectors identified in the IPCC report where mitigation technologies, policies and measures, constraints and opportunities exist, including energy supply, transport, buildings, industry, agriculture, forestry, and waste. International standards offer policy makers, industry and users the common tools they need to work together on tackling climate change. The three partner organizations also offer a comprehensive system in which nations and the private sector can participate to establish the priorities for tackling climate change in the years ahead. As such, they offer practical solutions with the potential to be used as part of any international agreement following on from the Kyoto Protocol. Sustainability will become a major impediment. Nobody can deny that the poor have a right to a better standard of living—be it access to safe water, better food and sanitation, and better education. The pressure to serve an additional 3 to 4 billion and at the same time protect the environment will focus attention on sustainability as never before. For that making choices that reduce greenhouse gas pollution, and preparing for the changes that are already underway, we can reduce risks from climate change. Our decisions today will shape the world our children and grandchildren will live in.

In order to understand the interplay of our consumption patterns and climate change, a National Seminar on "Sustainability Consumption, Climate Change and Consumers—Towards a Better Future" was organized by the Centre for Consumer Studies, Indian Institute of Public Administration, New Delhi. The present book is an outcome of the papers presented in the seminar and contains twenty one chapters written by eminent experts working in the area. They have discussed some of the most important and critical issues pertaining to climate change and global warming in India. It evaluates the measures taken to slower the pace of global warming which is hampering the environment leading to serious changes in climate all over and also leading to frequent natural disasters.

We are thankful to the Department of Consumer Affairs, Government of India for sponsoring the seminar. We would like to thank all the paper contributors for sending their revised papers in time. We are grateful to Shri S.N. Tripathi, Director, Indian Institute of Public Administration for his guidance and support. We would also like to thank all the members of the Centre for Consumer Studies for their support. We hope that the readers will find the book useful in understanding the critical issues pertaining to climate change and global warming in India and help in suggesting ways and means to come up with viable solutions for the same.

List of Contributors

1. **Wankhede, Archana,** Professor, Shankarlal Agrawal College of Management Studies, Nagpur University, Suryatola, Gondia, Maharashtra. (Res.) C/o Mr. Rokade, opp. Ashoka Vedio, Arihant Colony, Gondia, Maharashtra. (M): 9665405744, 9960775106, Ph. : 0712-22899471.
   Email: archanawankhede1@gmail.com
2. **Dey, Mithra,** Head, Department of Ecology and Environmental Science, Assam University, Silchar-788001, Assam. (M) : 09435073238 Email: mithradey@gmail.com
3. **Dey, Nikhil Bhusan,** Professor and Head (Retd.), Department of Commerce, School of Management Studies, Assam University, Silchar-788011.
   Email: nikhil.b.dey@gmail.com
4. **Chauhan, Sanchita,** Professor, University Institute of Chemical Engineering and Technology, Punjab University, Chadigarh-160014 India. H.No. 126-A, Sector-6, Panchkula, Haryana. (M): 9888814141. Email: sanchita@pu.ac.in
5. **Dhote, Meenakshi,** Professor, Department of Environmental Planning, School of Planning and Architecture, Indraprastha Estate, New Delhi-11002.
   (M): 91-11-9313508547
   Email: meenashidhote@gmail.com, m, dhote@spa.ac.in
6. **Lata, Kusum,** Associate Professor, Urban and Regional Planning, Indian Institute of Public Administration, IP Estate, Ring Road, New Delhi-11002. Ph: 011-23468442
   (M): 9911157618 Email: dr.kusmlata@gmail.com
7. **Slariya, Mohinder Kumar,** Associate Professor, Sociology, Government PG College, Chamba, Himachal Pradesh -176310. (M): 91-94180-41374 Email: mohinderslariya@gmail.com

8. **Mittal, Ms Radhika,** Research Scholar, Centre for Research on Social Inclusion, Macquarie University, Level 3, C5C Research Hub West, Sydney NSW 2109, Australia
Ph: +612 9850 1013 Email: radhika.mittal@mq.edu.au
9. **Bhatt, Dr. K.N.,** Professor, G. B. Pant Social Science Institute, Jhusi, Allahabad-211019. (M) : 09336899726
Email: knbhatt1@rediffmail.com
10. **Dutta, Mr. Ranjit,** Assistant Professor, Department of Botany, Madhavdev University, Lakhimpur, Assam. (M): 91-9435628642 Email: ran.jit2@yahoo.com
11. **Sarma, Prof. Sarada Kanta,** Professor (Retd.)
(M): +91-94353-03130 E-mail: sksarmah@sify.com
12. **Singh, Dr. Amit Kumar,** Assistant Professor, Indian Institute of Public Administration, IP Estate New Delhi-110002
(M): 9873922335, (O): 011-23468352
Email: amitsinghh@gmail.com
13. **Patro, Mr. K. Arup Kumar,** Programme Officer, Focus Humanitarian Assistance, India, An Affiliate of the Aga Khan Development Network (AKDN), 305, Maker Bhavan-3,21-New Marine Lines, Mumbai-400 020, Maharashtra.
Ph: 91-22-22017175/74, Fax: 91-22-22017173, (M) 91-9769511088 Email: arup.patro@focushumanitarian.org
14. **Mishra, Dr. Manish,** Research Scientist Selection Grade, Indian Institute of Forest Management, Nehrunagar, Bhopal, MP-462004 (M): 09993380768 Email:manishm@iifm.ac.in
15. **Parekh, Dr. Hetal,** Expert Zoologist, Gujarat Biodiversity Board, Gandhinagar, Gujarat (M): 8980298618
Email: het277@gmail./com
16. **Poriya, Mr. Paresh U.,** Assistant Professor, Zoology, Government Science College, Veraval, Bhakta Kavi Narsinh Mehta University, Junagarh (M): 9974953005
Email: pareshporiya@gmail.com
17. **Sharma, Prof. Vinod K.** Professor (Retd.), Disaster Management & Environment, Indian Institute of Public Aministration, I.P. Estate, Ring Road, New Delhi-110002
(M): +91 98189-61977
Email: profvinod@gmail.com
18. **Perumal, Dr. S. Arumuga,** Professor and Head, Department of Computer Sicence, S.T. Hindu College, Nagercoil-629 002, Tamil Nadu. (M): 91-9443402050 Email: visvenk@yahoo.co.in

19. **Simon, Mr. Sabu. M.** Senior Programme Officer, Peermade Development Society, P.B. No. 11, Peeermade, District-Idukki, Kerala-685 537. Ph: 09895032198 E-mail-sabumsimon@gmail.com
20. **Susai, Mr. Diraviaraj Maria,** Secretary, Y-NEEW, No. 28, Second Street, Sasthiri Nagar, Dindigul- 6240031, Tamil Nadu (M): 09842780640 E-mail: diravigreenearth@gmail.com
21. **Sathseesh, Dr. E.K.,** Professor & Dean, Department of Commerce and Management Studies, University of Calicut, Kerala- 673 635. (M): 09400441150 Email: eksatheesh@gmail.com, satheeshek@yahoo.co.in
22. **Malodia, Dr. Suresh,** Assistant Professor, Mica Ahemdabad. (M): 91-9456154543 Email: smalodia@jimindore.ac.in, surmalodia@ gmail.com, Suresh.malodia@micamail.in
23. **Singh, Mr. Anirudh Prasad,** Ph. D. Scholar, Academy of Third World Studies, Jamia Millia Islamia, New Delhi (M): 9891333536 Email: personal1933@yahoo.com
24. **Singh, Ms Jyoti,** Research Scholar, V.M. University (Education Department), Salem, (Res.) NH-47, Sankari Main Road Ariyanoor, S. Veerapandi (PO) Salem-636 308, Tamil Nadu.
25. **Kumar, Dr. K. J. Anandha,** Scientist, Ministry of Jal Shakti, MTNL Building, CGO Complex, Lodhi Road, New Delhi-110 003 (M): 09868165858, Email: kajanand@gmail.com
26. **Mall, Dr. R. K.,** Professor, Banaras Hindu University (M): 8765447799 Email: mall_raj@rediffmail.com
27. **Soundararajan, Dr. K.,** Professor in Business Administration, DDE, Annamalai University, A. Nagar, Tamil Nadu-608002 (Res.) No. 31-33, East Car Street, ARN Apartment, Chidambaram-608 001 (Tamil Nadu) Ph: 098940 82552 Email: spknrajan@gmail.com
28. **Ramanathan, Prof. A. L.,** Professor, Lab No. 214, School of Environmental Sciences, Jawaharlal Nehru University, New Delhi-110067. (M)-09810689243 Email: alrjnu@gmail.com
29. **Singh, Shri Virendra Bahadur,** Research Scholar, Lab No. 214, Glacier Research Group, School of Environmental Sciences, Jawaharlal Nehru University, New Delhi-110067 (M): 09999310454. Email: virendrasinghjnu@gmail.com
30. **Jose, Shri P.G.,** School of Environmental Sciences, Jawaharlal Nehru University, New Delhi.

Contents

Introduction

Suresh Mishra and *Mamta Pathania*

Introduction

Our lives are connected to the climate. Human societies have adapted to the relatively stable climate we have enjoyed since the last ice age which ended several thousand years ago. A warming climate will bring changes that can affect our water supplies, agriculture, power and transportation systems, the natural environment, and even our own health and safety. Some changes to the climate are unavoidable. Carbon dioxide can stay in the atmosphere for nearly a century, so Earth will continue to warm in the coming decades. The warmer it gets, the greater the risk for more severe changes to the climate and Earth's system. Although it is difficult to predict the exact impacts of climate change, what is clear is that the climate we are accustomed to is no longer a reliable guide for what to expect in the future. Global warming does not mean that our planet will turn into a tropical island. Instead, it means that the weather we are used to, will become very unpredictable. Some places may experience droughts, while others will be flooded. Some places will become unbearably hot, while others will be covered by blizzards. As the ice caps melt, the level of the world's oceans will rise, permanently flooding many of the coastal areas where people are living now.

Earth's average temperature has risen by 1.4°F over the past century, and is projected to rise another 2 to 11.5°F over the next hundred years. Small changes in the average temperature of the planet can translate to large and potentially dangerous shifts in climate and weather. Rising global temperatures have been accompanied by changes in weather and climate. Many places have seen changes in rainfall, resulting in more floods, droughts, or intense rain, as well as more frequent and severe heat waves. The planet's oceans and glaciers have also experienced some big changes—oceans are warming and becoming more acidic, ice caps are melting, and sea levels are rising. As these and other changes become more pronounced in the coming decades, they will likely present challenges to our society and our environment.[1]

The world is facing a critical challenge. Increasing greenhouse gas emissions are raising the earth's average temperature. As a result, dramatic climate change is forecast and global scientific opinion predicts enormous developmental, economic, social and environmental stresses on our planet. Leading climate change experts have put forward a series of practical solutions to tackle climate change. These solutions include the technical standards published by the world's three leading international standardization organizations: the International Electrotechnical Commission (IEC), the International Organization for Standardization (ISO) and the International Telecommunication Union (ITU). In its groundbreaking report published in 2007, the Intergovernmental Panel on Climate Change (IPCC) cited technical standards, like those published by the IEC, ISO and ITU, as a means of mitigating climate change now, while offering the potential to reduce its effects in the future as new technologies are developed and mature. The three global organizations are coordinating their work to ensure that government, business and society are provided with the necessary tools to help combat global climate change and to support the reduction in greenhouse gas emissions by increasing energy efficiency, while facilitating sustainable development.

The standards produced by the three organizations reach across all sectors identified in the IPCC report where mitigation technologies, policies and measures, constraints and opportunities exist, including energy supply, transport, buildings, industry, agriculture, forestry, and waste. The IEC, ISO and ITU offer a system of standardization whose output includes standards for the following aspects of the fight against climate change:

- Monitoring and measurement of greenhouse gas emissions.
- Measuring the carbon footprint of networks and products
- Designing and building energy efficient homes and workplaces.
- Benchmarking for good practices including environmental and energy efficiency labeling.
- Promoting good practice for environmental management and design, and for energy management.
- Disseminating innovative technologies that promise to help reduce the effects of climate change.
- Fostering the introduction of new energy-efficient technologies and services.

International standards offer policy-makers, industry and users the common tools they need to work together on tackling climate change. The three partner organizations also offer a comprehensive system in which nations and the private sector can participate to establish the priorities for tackling climate change in the years ahead. As such, they offer practical solutions with the potential to be used as part of any international agreement following on from the Kyoto Protocol.[2]

The Kyoto Protocol is an international agreement linked to the United Nations Framework Convention on Climate Change, which commits its Parties by setting internationally binding emission reduction targets. Recognizing that developed countries are principally responsible for the current high levels of GHG emissions in the atmosphere as a result of more than 150 years of industrial activity, the Protocol places a heavier burden on developed nations under the principle of "common but differentiated responsibilities." The Kyoto Protocol was adopted in Kyoto, Japan, on 11 December, 1997 and entered into force on 16 February, 2005. The detailed rules for the implementation of the Protocol were adopted at COP 7 in Marrakesh, Morocco, in 2001, and are referred to as the "Marrakesh Accords." Its first commitment period started in 2008 and ended in 2012. Under the Protocol, countries must meet their targets primarily through national measures.

However, the Protocol also offers them an additional means to meet their targets by way of three market-based mechanisms:

(i) International Emissions Trading (IET);
(ii) Clean Development Mechanism (CDM); and
(iii) Joint Implementation (JI).

The mechanisms help to stimulate green investment and help Parties meet their emission targets in a cost-effective way. In Doha, Qatar, on 8 December, 2012, the "Doha Amendment to the Kyoto Protocol" was adopted. The amendment includes:

- New commitments for Annex I Parties to the Kyoto Protocol who agreed to take on commitments in a second commitment period from 1 January, 2013 to 31 December, 2020;
- A revised list of Green House Gases (GHG) to be reported on by Parties in the second commitment period; and
- Amendments to several articles of the Kyoto Protocol which specifically referenced issues pertaining to the first commitment period and which needed to be updated for the second commitment period. During the first commitment period, 37 industrialized countries and the European Community committed to reduce GHG emissions to an average of five per cent against 1990 levels. During the second commitment period, Parties committed to reduce GHG emissions by at least 18 per cent below 1990 levels in the eight-year period from 2013 to 2020.[3]

Role of Different Sectors in Mitigatory Measures of Climate Change

High Powered Committee (HPC) adopted a strong process oriented and participatory approach involving different sectors of the society like NGOs, Media, Police and Security Forces, Communications and Governments including Administrative Training Institutes (ATIs) at all levels, etc. as against an academic and technical approach (HPC Report, 2001).

India ratified the Kyoto Protocol in 2002. Being a developing country it was not required to cut emissions, but it has developed certain mechanisms to reduce emissions. The primary legislation in India is as follows:

- National Hydro Policy, 1998—Outlined strategies for hydro power potential.
- Energy Conservation Act, 2001—Established Bureau of Energy Efficiency.

A National Bio Diesel Policy is also being considered to encourage use of Biodiesel.

The Central Government especially National Institute of Disaster Management (NIDM), Ministry of Home Affairs and all ATIs of State Governments through their Natural Disaster Management (NDM) faculties may start the training programmes on climate change by incorporating the contribution of different research centres working in the country and abroad. They may also conduct the training programmes on environmental Laws for the officers of Central Pollution Control Board (CPCB), State Pollution Control Board (SPCB) and Public Prosecutors of States and UTs as they may be helpful in exploring the knowledge of various Acts of Environmental Protection, Water and Air Pollutions, Wild life, Fisheries and Forests, etc. among the people and force them to follow these Acts strictly.

The role of the Electronic Media has during recent times emerged as a major component in the field of global climate change. At the same time, the role of the Print Media cannot be taken on a lower level of importance because it continues to be the medium of mass media in many parts of the Indian society still unreached by the electronic media. The Educational Bodies apart from School to University level have been implementing educational and awareness programmes on environment including climate change.

As we all know India is an agriculture based economy and even little change in the duration and amount of rainfall will be having greater impact on agriculture. With the climate variability agriculture production in India (like many other developing countries) would be seriously compromised and productivity of some of the crops like wheat, rice and pulses will decline considerably which will adversely affect the self sufficiency in food production. As India is highly vulnerable for number of natural disasters including droughts and floods. The country's 58 per cent sown area is vulnerable for droughts. Climate change could result in higher temperatures and droughts that are more frequent, heavy floods in the Indo-Gangetic plains, un-seasonal rainfall posing a threat to crops nearing maturity, and a rise in sea level along the coastal areas. This may lead to food insecurity and further enhance poverty related endemic hunger.

However, we can reduce the risks we will face from climate change. Sustainability will emerge as a source of innovations at the Bottom of the Pyramid: The planet is under stress—be it access to water, deforestation, pollution, or greenhouse gas emissions. What if we added an additional 3 to 4 billion people as micro producers and consumers to the current few billions? Sustainability will become a major impediment. Nobody can deny that the poor have a right to a better standard of living—be it access to safe water, better food and sanitation, and better education. The pressure to serve an additional 3 to 4 billion and at the same time protect the environment will focus attention on sustainability as never before. For that making choices that reduce greenhouse gas pollution, and preparing for the changes that are already underway, we can reduce risks from climate change. Our decisions today will shape the world our children and grandchildren will live in.

About the Book

The book is an outcome of a National seminar on "Sustainability Consumption, Climate Change and Consumers—Towards a Better Future", organized by the Centre for Consumer Studies, Indian Institute of Public Administration, New Delhi. The chapters contributed by well known experts discuss some of the most important and critical issues pertaining to climate change and global warming in India. It evaluates the measures taken to slower the pace of global warming which is hampering the environment leading to serious changes in climate all-over and also leading to frequent natural disasters. The case studies identify the causes of the increase in global warming and suggests future course of action in this direction with certain remedial measures.

Climate Change in India

The chapter on "Climate Change—A Global Concern" by *Archana Wankhede,* discusses the changes in the climate describing the statistical distribution of weather over periods of time that range from decades to millions of years. Defining the term in recent usage, especially in the context of environmental policy, climate change usually refers to changes in modern climate. It may be called as anthropogenic climate change. As climate change is a global issue according to her due to rapid industrialization and advancement of technologies there is lot of pollution and this is the main reason for climate change. Because of which there is a negative effect on the environment. According to the author, the anthropogenic factors i.e. the increase in CO_2 levels are due to emissions from fossil fuel combustion, followed by aerosols (particulate matter in the atmosphere) and cement manufacture. Other factors, including land use pattern, ozone depletion, agriculture and deforestation, are also of concern in the roles they play—both separately and in conjunction with other factors which affect climate, micro-climate, and measures of climate variables.

This is directly related to economy of a nation also. The global recession and economic despair temporarily knocked the climate change issue off the top line agenda, but as the recession is now beginning to recede, the Copenhagen Summit has pushed this important issue to the forefront again. She highlights globally, the majority of consumers still believe that the main responsibility for solving climate change should lie with their governments. In October 2009, 36 per cent of global consumers said that governments should restrict companies' emissions of carbon dioxide and other pollution, closely followed by 34 per cent who said there should be major government led initiatives for research into scientific and technological solutions such as low emission cars, houses and renewable energy. She concludes by saying about one in three global consumers also believe there should be government incentives to individuals for good, less or non-polluting behaviour and that the population should recycle waste whenever possible.

Dr. Sanchita Chauhan in her chapter, "Role of Catalytic Converters in Controlling Pollution" describes that climate change is primarily due to unchecked release of pollutants to the atmosphere from various sources. Sustainable development requires not only judicious use of resources but also use of advanced technologies to combat the release of pollution. The need for mobility in lifestyle has led to a drastic increase in air pollution. Combustion of fuels in automobiles leads to formation of many harmful pollutants like carbon monoxide, unburnt hydrocarbons and oxides of nitrogen. As these pollutants are released near the ground level, they prove to be more harmful to the health of human beings than the industrial pollutants. The author further mentions that complete reduction of harmful pollutants emitted by vehicles is a complex procedure as it includes numerous players like automobile manufacturers, government agencies, catalyst suppliers, petroleum refiners, and fuel-additive suppliers, among others.

However Catalytic Converters attached in automobiles are considered as the most cost-effective means to achieve drastic reduction of pollutants present in exhaust gases. In this chapter, the advancements in the converters brought over the period of time, methods and catalytic models employed for the reduction of exhaust pollutants have been discussed. Dr Sanchita opines that although they have been extensively studied, still there are some areas that need to be further explored like the cold start period, during which the hydrocarbon emissions are relatively high. Mathematical modeling has proved to be a good way of analysing the physicochemical processes occurring in the converter. Modeling is preferred as compared to experimental testing, the later being very expensive and time consuming. Modeling can be applied for design purposes and improving the existing converter control strategies. Difficulties encountered during modeling are due to complexities in reaction schemes and the rate expressions to be used for different catalytic formulations.

Prof. Meenakshi Dhote and Dr Kusum Lata in their chapter, "'Resource Management for Climate Resilience in Urban Areas" mention that with urban areas growing at an accelerating pace, there is an urgent need to be disaggregated and their correlation with climate parameters has to be understood by policy makers, planners, engineers, environmentalists, etc. so as to ponder and suggest pragmatic solutions. Land use pattern of urban areas is the result of an exercise that takes into account the environmental, technological, social, economic, political issues of a settlement and translates it according to a future vision onto the ground. Thus, the urban land use planning exercise is a critical aspect when it comes to controlling environmental conditions and impacts of any kind. The chapter starts with appreciation of natural resources in urban areas followed by analysis of the study of the land use patterns of a metropolitan city, emphasizing the open space and water bodies that exists therein. The natural resources in the urban areas and its role in climate control are explained to understand its value. Efforts of human resources (citizens, organisations, agencies, etc.) need to be streamlined for effective climate change resilience.

The authors recommend an integrated approach involving control of emissions, sequestration of gases with judicious allocation of green spaces, coordination in various efforts of human resources needs to be framed. The approach varies at the level of a building, neighbourhood, planning zone, city and a metropolitan region. Methods of assessment and conservation have been discussed. Finally, an attempt has been made through this chapter to outline the elements of land use strategy for natural resource conservation along with institutional framework with proper O&M system to address the climate change in urban areas.

Ms Radhika Mittal in her chapter, "The Environmental Impact of Unsustainable Food Practices: A Perspective on Food and Ecological Disaster" discusses the anthropogenic greenhouse gas emissions that contribute to climate change, food consumption patterns which have only recently been addressed. Through this review, the researcher looks at the startling data on global food production and consumption practices and its related environmental impact. From extensive animal farming to poor agricultural practices, from food miles through food transportation to the shift to processed, packaged and manufactured foods, food practices have grown increasingly unsustainable. For instance, livestock and its byproducts alone account for at least 51 per cent of annual worldwide greenhouse gas emissions (Goodland and Anhang, 2009). The chapter intends to articulate the impact of human food choices in an industrialized, globalised world that faces an environmental crisis. This summary will serve as an incentive to reconsider our lifestyle decisions and be attuned to alternatives. In this light, a country like India, where the emphasis has traditionally been on an eco-centric way of life and the concept of simple, sustainable living was readily embraced and practiced, has the potential to show the way forward.

Dr Amit Singh in his chapter, "Fresh Water Resources in

India: Issues and Challenges", says that fresh water is the most fundamental of finite resources. It has no substitutes for most of its use. But fresh water resources are dwindling on becoming contaminated throughout the world. The problem is particularly serious in developing countries, where agricultural production as well as productivity is very low with a fast growing population. Chronic or acute water shortage is increasingly common in many of these countries. At present more that 80 per cent of the world's population is living in developing countries and large section of them do not have safe drinking water supply. However, water is not only essential to meet drinking requirements, but is also essential to percolate economic development at grass root levels through agricultural and industrial development itself.

The chapter further highlights that right now, we are already consuming 55 per cent of the annual available fresh water resources throughout the world and it is predicted that by 2025 we will consume 70 per cent of it. This estimate reflects only the impact of population growth on fresh water resources; however the growing discharge of domestic sewerage, industrial waste water and discharge of contaminated agricultural water into fresh water bodies is severely damaging the fresh water. In developing countries, 90-95 per cent of sewage and 70 per cent of industrial waste are dumped into surface waters thus polluting the water supply. India is no exception to that. Fresh water is of vital importance to India's future. Historically it has been an important component of its culture, economy, early transportation systems and resource development. It further mentions that as India is an agriculture based country where more than 65 per cent of the population still depends on agricultural and allied activity for their livelihood. Estimates for India's population growth indicates that by the year 2025 it will become 1.35 billion and by the year 2050 it will become 1.65 billion, therefore, the demand for fresh water will definitely increase manifolds for drinking water, industrial manufacturing, power generation and agricultural irrigation. Against this background, this research chapter tries to assess the impact of population growth, agricultural practices, urban water demand and waste generation, industrial water pollution on fresh water resources in India.

Dr K. Arup Kumar Patro in his chapter, "Participatory Sustainable Mangrove Conservation and Management of Coastal

Wetlands", describes that although mangrove ecosystems have tremendous value for coastal communities and associated species, they are being destroyed at alarming rates. Over the last 50 years, about one-third of the world's mangrove forests have been lost (Alongi, 2002). Human threats to mangroves include the overexploitation of forest resources by local communities, conversion into large scale development such as agriculture, forestry, salt extraction, urban development and infrastructure, and diversion of fresh water for irrigation (UNEP, 1994). It clearly brings to light that greatest human threat to mangroves is the establishment of shrimp aquaculture ponds. Because mangroves are often viewed as wastelands, many developing countries are replacing these forests with agricultural land and/or shrimp aquaculture production (Franks and Falconer, 1999). Shrimp aquaculture accounts for the loss of 20 to 50 per cent of mangroves worldwide (Primavera, 1997).

In addition to these anthropogenic threats, mangroves are also threatened by the impact of global climate change. Global climate change and concomitant effects such as changes in temperature and CO_2, altered precipitation patterns, storminess, and eustatic sea level rise as observed over recent decades, are due primarily to anthropogenic activities. Most of the observed warming over the last 50 years is attributed to an increase in greenhouse gas concentrations in the atmosphere (Houghton *et al.*, 2001). The project "Coastal Wetlands: Mangrove Conservation and Management" M.S. Swaminathan Research Foundation has been implemented with support from ICEF in four States i.e. Tamil Nadu, Andhra Pradesh, Orissa and West Bengal to enhance national capacity and national action in the conservation and sustainable management of the mangrove wetlands of the east coast of India.

Dr. Anirudh Prasad Singh and Dr. Jyoti Singh in their chapter, "The Clean Development Mechanism (CDM) and Carbon Trading" highlight that the carbon trading is one of the fastest growing financial markets in the world. It is the most visible result of early regulatory efforts to mitigate climate change and grew out of the Kyoto Protocol, which was adopted in 1997. We are in debt to Carbon Dioxide (CO_2) and other green house gases for our presence on earth. As they help in stabilizing temperatures to levels sustainable for organic life—by what is known as green house effect.

In modern times burning of fossil fuels like coal, oil, and natural gas combined with fast deforestation has led to unprecedented level of green house gas emission. Here came into existence the concept of Carbon Emission Trading developed during the Kyoto Protocol in 1997. To help countries achieve that goal, it established the Clean Development Mechanism (CDM), which promotes sustainable development in developing countries. Developed countries can implement projects that reduce emissions or remove carbon from the atmosphere in other developing countries in lieu of CERs (Certified Emission Reductions). In recent years, significant global governmental involvement in the stabilization of the concentration of atmospheric greenhouse gases has facilitated tremendous growth in the market for carbon emissions trading. India offers a large potential for CDM because of its inherent dependence on fossil fuels for development. So countries with relatively low abatement and transaction costs like India are a major attraction for CDM projects. The purpose of this research chapter is to get an overview on Carbon Credit and its related concepts, CDM Mechanism and process and to understand the eligibility criteria for Carbon Credits to help mitigate the increase of CO_2 in the atmosphere.

Dr. E.K. Satheesh in his chapter, "Sustainable Agricultural Production and Consumption Based on Total Quality Management" describes that rising food prices is one of the greatest concerns of societies all-over the world. This situation is considered to be a setback to the poverty alleviation programmes and overall health improvement scenario of the people living in underdeveloped countries. India is the second largest food producer in the world after China. Indian food market is estimated to be at US $ 182 billion and the retail food sector in the year 2008 was US$ 70 billion. This is expected to grow at US $150 billion by 2025. The chapter further describes that the Union Government has already taken steps to prevent food shortage caused by the ongoing monsoon failure.

At the time of independence, the contribution of agriculture to GDP was 70 per cent and employing nearly 85 per cent of the total population. Over a period of six decades, the contribution of agriculture to GDP was drastically reduced to 18 per cent. But the percentage of employment in agriculture more or less remains the same (70%). India expects a GDP growth of 8 per cent per annum. This indicates that the agricultural sector should grow at the rate of 4 per cent, but at present it is growing only at 2 per cent. Thus, there should be proper forward integration aiming at the value addition of agricultural sector. The food processing sector should be given further impetus so as to make the agro processing industry more sustainable and quality oriented.

Prof. A.L. Ramanathan, Mr. Virendra Bahadur Singh and Mr. P.G. Jose through their chapter", Characteristics of Raktvarn Nala of Gangotri glacier, Garhwal Himalaya, India" focus on estimating the meltwater characteristics in terms of weathering and different chemical compounds of the Raktvarn nala of Gangotri glacier. In the melt water the presence of cations varied as $c(Ca^{2+})> c(Mg^{2+})> c(K^{+})> c(Na^{+})$, while the order of concentration of anions has been $c(SO4^{2-})> c(HCO_3^-)> c(Cl^-)>c\ (NO_3^-)$ in year 2007. It highlights that the average value of EC and pH of Raktvarn nala of Gangotri glacier melt water was found to be 214.4 μs/cm (174.4-257.8 μs/cm) and 5.8 (5.5-6.1) respectively. The relative high contribution of $Ca^{2+}+Mg^{2+}$ to the total cations (TZ^+), the high $Ca^{2+}+Mg^{2+}/Na^{+}+K^{+}$ ratio (6.71) and the low $(Na^{+}+K^{+})/TZ^{+}$ ratio (0.13) indicate that carbonate weathering is the major source of solute in the melt water of Raktvarn nala of Gangotri glacier.

It also mentions low concentration of NO_3^- at the sampling site was reported which indicated that nitrate is derived exclusively from the atmosphere, associated with the acidic nitrate aerosols. According to Gibbes diagram Raktvarn nala of Gangotri glacier meltwater was controlled by weathering process. Thermodynamic stability diagram for Na, K, Ca, Mg system shows that meltwater of Raktvarn nala of Gangotri glacier is in the range of stability field of Kaolinite.

Dr. K.J. Anandha Kumar and Dr. R.K. Mall in their chapter, "Groundwater Resource Depletion and Need for Development of Sustainable Communities" describe that groundwater is a replenishable and considered to be a finite resource. Rainfall is the principal source of its recharge, in general. Groundwater comprises of two parts namely dynamic, in the zone of water table fluctuation and static resources, below this zone, which usually remains perennially saturated. It further highlights that with the manifold rise in population, rapid urbanization, industrialization, development of technologies in agriculture, the demand of water has grown over the recent past, especially in the last two decades. This associated with the technological developments in the groundwater drilling sector has promoted withdrawal of groundwater, as it can be withdrawn as and when required, at place of requirement and easily accessible, free from contamination unlike the surface water sources. The groundwater development of the country stands at 58 per cent for the country as a whole, but there are more than 1615 assessment units which utilize more groundwater than which is safer for environment. Out of these there are 839 units where the withdrawal is more than what is being recharged through the rainwater. Further to add to woes, the climate change also poses a threat to groundwater resources as the decline in rainfall intensity over space and time and rainfall pattern will have direct effect on groundwater recharge and its utilization.

Thus, the finite groundwater resource has been under threat from quantity decline and quality deterioration due to several reasons. There is thus an urgent need for awareness among communities and involvement of communities for conservation, augmentation and management of groundwater resource. It stresses that the Government policy should aim at encouraging the involvement of communities for development and management of groundwater as the regulation of groundwater will be an impossible task for the Governmental agencies, without involving the communities as seen through experience. Good practices are to be promoted with incentives so that such practices spread to other parts of the country, thus creating a sustainable community which can manage their water resources at local level, in a sustainable manner.

Ms H.M. Parekh and Mr. P.U. Poriya in their chapter, "Ocean Acidification: Great Risk to Oceans by Climate Change" describe that global warming is a major issue and the main cause of global warming is release of carbon dioxide, primarily through the burning of fossil fuels and the burning of biofuels. Over the time, the oceans have helped this problem by absorbing excess carbon dioxide. According to NOAA, the oceans have absorbed nearly half of the CO_2 emissions generated over the past 200 years. The chapter highlights that ocean acidification is the name given to the ongoing decrease in the pH of the oceans of the world, caused by their uptake of anthropogenic carbon dioxide from the atmosphere. Dissolving CO_2 in seawater increases the hydrogen ion concentration (pH) in the oceans, and thus decreases pH of ocean. Further ocean acidification from CO_2 emissions causes substantial irreversible damage to oceanic ecosystems. Seas and oceans absorb approximately one quarter of the carbon dioxide emitted to the atmosphere from the burning of fossil fuels, deforestation, and other anthropogenic activities.

As more and more carbon has been emitted into the atmosphere, the oceans have absorbed greater amounts at increasingly rapid rates. Without this level of absorption by the oceans, atmospheric CO_2 levels would be significantly higher than at present and the effects of global climate change would be more marked. This can have drastic consequences on corals and other marine life, with cascading impacts on the fishing and tourism industries. The study shows that increasing ocean acidification will mean that by 2100 some 70 per cent of cold water corals, a key refuge and feeding ground for commercial species of fish, will be exposed to corrosive waters. Lowering seawater pH with CO_2 affects biological processes, such as photosynthesis, nutrient acquisition, growth, reproduction, and individual survival.

Prof. Vinod K. Sharma in his chapter, "Climate Change—Issues and Challenges" describes about the reports of Intergovernmental Panel on Climate Change (IPCC) which have made it clear that climate change is happening largely because of human activities. This is also known that impact of global warming will be more visible in the developing countries. India is an agriculture based economy and even little change in the duration and/or amount of rainfall will be having greater impact on agriculture. With the climate variability agriculture production in India (like many other developing countries) would be seriously compromised and productivity of some of the crops like wheat, rice and pulses will decline considerably which will adversely affect the self-sufficiency in food production.

As India is highly vulnerable for number of natural disasters including droughts and floods, country's 58 per cent sown area is vulnerable for droughts. Climate change could result in higher temperatures and droughts that are more frequent, heavy floods in the Indo-Gangetic plains, un-seasonal rainfall posing a threat to crops nearing maturity, and a rise in sea level along the coastal areas. This may lead to food insecurity and further enhance poverty related endemic hunger. The chapter highlights that India has prepared a National action plan on Climate Change which was given to the concerned ministries and departments to prepare their own plan for climate change mitigation and adaptation. A number of states have also taken adequate measures and prepared state specific plans. The present chapter discusses the Indian national policy, action plan and current major issues and challenges. The chapter also highlights the existing schemes like Rural Self Employment Programme, Comprehensive Rural Employment Scheme, Prime Minister's Rural Road Programme, National Rural Health Mission, Accelerated Rural Water Supply Programme, and Sustainability of Dry Land/Rainfed Farming System may be used for short and long term mitigation of the impact of climate change.

Dr. S. Arumuga Perumal in his chapter, "Green Computing—Best Practices to Save Environment" describes about green computing which is the use of practices that reduce the environmental impact of usage of computers and the computer manufacturer to save the country from natural disaster. The everyday use of computers and replacement of existing computers with new configuration can be hazardous to the environment. Computer systems are made of chemicals and materials that cannot be disposed of safely in the environment and the electronic devices used in computers require a good amount of energy. The aim of this chapter is to reduce the environmental impact of replacement of the old computer and safeguard one's private information that is stored in storage devices. The files that users think they delete from the old hard drive can be recovered and used again. If one simply throws a computer away, we seriously compromise with all the confidential files, trade secrets and proprietary data. If the discarded computer falls into the hands of a person who knows how to access the hard drive, then he can freely retrieve the confidential data and use it without a trace, such as one's name, birth date, social security number or credit card numbers.

So we are in a position to keep the computer users knowledge base up-to-date on any new dangers that they should be cautious about. The development of Information Technology leads to the remarkable growth in all the fields, but the users don't know what is happening in the darkness of the internet world. The author further recommends in order to reduce the environmental impact of our old computer we can donate it to a recycling centre instead of throwing it in the trash. The best thing to do with the hard disk drive is physically destroy it. Green computing is a great way to help the environment from the worthy cause, but we should be wise about protecting any sensitive information when we recycle.

Dr. P. Chandrasekaran and Dr. R.B. Gnana Jothi in their chapter, "Integrated Bio-Organic Farming Technology (IBFT)—A Promising Tool for Poverty Alleviation" highlight that realizing the problems of population explosion, depletion of natural resources and damages to nature, it is high time to integrate biological disciplines and to promote biology related indigenous technology through indigenous and modern means appropriately with the students of biology and biology integrated disciplines. This chapter deals with an environment friendly, integrated bio-organic farming programme which links various disciplines such as Aquaculture, Horticulture, Floriculture, Vermi-Composting, Apiculture, Rabbit farming, Dairy farming, Mushroom farming etc. by using the output of one system as the input for another system. It aims at promoting IBFT skills with and through the students of biology and biology integrated discipline to the rural farmers and rural unemployed youth, leading possibly towards alleviation of regional poverty. The author through this working pilot model project of PG and Research Department of Biotechnology, Vivekananda College, Madurai, involves the promotion of awareness with rural people on right and optimal integration of resources and practices towards right and maximum productivity with minimal risks to the consumer and earth.

It says permaculture movement is one which involves the integration of various main systems and related subsystems. This strives for the harmonious integration of human dwellings, microclimate, annual and perennial plants, animals, soils, and water into stable, productive communities. Integrated farming and "ecological engineering" are the practical examples used to describe permaculture, with "cultivated ecology". Permaculture principles focus on thoughtful designs for small-scale intensive systems which are labour efficient and which use biological resources instead of fossil fuels. Designs stress on ecological connections with closed energy and material loops integrating the resources optimally for maximal sustainable production. The core of permaculture is, design the working relationships and connections between all of them. Each component in a system performs multiple functions, and each function is supported by many elements. Permaculture is not limited to plant and animal agriculture, but also includes community planning and development, use of appropriate technologies (coupled with an adjustment of lifestyle), and adoption of concepts and philosophies that are both earth-based and people-centred, as Mahatma Gandhi had referred, Earth is for everybody's need and not for everybody's greed.

Dr. Suresh Malodia through his chapter, "A Study of Consumer's Attitude and Behaviour Toward's a Sustainable Energy System in India" highlights conservation is the first law of natural world. Inefficient use of energy results as a wastage and has a gripping environmental impact with regional and global implications. The primary footstep towards a sustainable energy system is thoughtful use of the existing possessions. The largest growth is anticipated to come from the budding economies of Asia. As a matter of fact, the estimated demand in this region is expected to double in coming years. Where will this oil come from? What are its implications for the future? And, how will this affect the energy needs of India? These are the questions which needs answers.

The author clearly brings to light that the challenge in creating a sustainable energy system for the future lies in two broad segments. The first is technology advancements in both conventional and non-conventional methods of energy generation, and mobilizing an ample financial support mechanism to execute the same; the second segment is developing a favourable attitude towards energy conservation. The latter is quite vital in gauging the level of our perception towards the need to conserve. This chapter thus studies the level of awareness towards energy conservation methods and importance given to energy conservation by industries as well as individuals in domestic segment. In addition to this, multivariate analysis is used to determine the underlying dimensions of attitudes towards energy conservation.

Dr. N.P. Singh in his chapter, "Impact of Climate Change and Consumer on Urbanisation" describes that climate change has emerged as one of the most important issues of urbanization in India. Multi cropping system changed to mono-cropping system due to climate change in rural areas. Traditionally people in villages had different types of crop farming which was supported by climate. There was sustainable cropping system in *Kharif* season. Madua, Kodo, Sawa, Arhar, Urad, Maize, short term paddy, long term paddy, small paddy plant and big paddy plant, etc. were the major crops in *Kharif* season in rural areas. Some of the crops among the above were able to grow in either flood or drought situations. Likewise in *Rabi* season, there were variety of crops in the Gangetic plain. People were able to sustain in all critical situations due to crop production. Climate change forced people to be dependent on wheat or paddy production. They required money for other groceries. Therefore, they were forced to migrate to urban areas for sustainability and livelihood in either of the situations.

There were different types of fruits in Gangetic plain which were subsidiary source of food and high calorie intake as well as economy. Approximately, eighty per cent fruit trees have been declined due to climate change. Sisam, Mahua and Neem etc. were natural trees. They were considered as life supporting trees. They support fuel, house construction, furniture as well as environment. Sisam tree was an asset for girl's marriage. Sisam and Mahua trees died during last one and a half decade due to climate change. This study is based on primary survey, using random sampling technique. Primary survey was conducted among the migrants in Delhi from Saran district of Bihar and justified by field visits.

Climate Change: Case Studies

Dr. Mithra Dey and Prof. Nikhil Bhusan Dey in their chapter "Impact of Climate Change on Food Security: A Study on Water Resource Management in Cachar District, Assam, India" highlight climate change as one of the most serious problems the human civilization is confronting in the recent years. Climate change is evident in the form of global warming, melting of glaciers, rise in sea level, unprecedented floods, drought or even cloud blast, rise in incidence of diseases etc. They opine India as a whole and Assam in particular has an agriculture based economy with most of the socio-economic condition of the population being determined by agriculture and its allied activities. Agriculture in Assam is largely dependent on rainfall and natural water resources, as irrigation facilities are insufficient and inadequate. Cachar is one of the three districts of Barak Valley region of Southern Assam, having a distinct topography with river Barak being the major river. River Barak and its tributaries have a meandering pathway and have unique fresh water flood plain water bodies (beels and haors), ponds, and lakes. There are small hillocks which are primarily used for tea cultivation. People in Assam are largely dependent on rice and fish which forms the staple diet. Paddy is the most dominant crop in the district along with vegetables, oilseeds, arecanut, banana, potato, sugarcane etc. The diverse water bodies serve as site of fish culture and also natural source of fish.

However, fish production within the district is not sufficient to meet the demands and fish has to be imported from other states. Paddy cultivation and fish culture is not intensive. They through their chapter clearly show that agriculture is highly dependent on rainfall in spite of the large natural water resource available. Further they bring out that the rainfall pattern has undergone considerable change in the recent years and a decline in annual rainfall is observed. This is accompanied by sudden heavy rains leading to inundation of low-lying areas which cause floods and hampers both crop as well as fish cultivation.

Dr. Mohinder Slariya in his chapter, "Impacts of Dams on Climatic Conditions and Agriculture of Hills—A Case Study of Chamera-I Hydroelectric Power Project in Ravi Basin of Himachal Pradesh" describes how installation of power projects have had serious impacts on the lives, livelihood, culture and spiritual existence of indigenous, tribal and illiterate people, and also on the physical environmental conditions and biodiversity. As far as the dam related developmental activities in Himachal Himalayas are concerned, they got momentum during 5th Five Year Plan and at present there are more than 400 power projects in five perennial river basins spread over 55,673 sq. km. geographical area. Every basin is being excessively exploited which has a considerable impact on environment, threatening the lives and livelihood of the mountain inhabitants.

The present chapter is based on the primary research carried out in Chamera-I (540 MW) catchment area under UGC sponsored research project in Ravi basin of the state. 29 km long reservoir in upstream and 12 km dry area in the downstream have unintentionally produced weather and climate changes on a larger scale and affected/threatening the existing land use pattern and sources of livelihood by interfering in the ecosystem. In the present chapter the responsibility of hydroelectric power projects for threatening the livelihood and climatic conditions is analyzed, which is based on original micro field research carried out in the lower Himalayan Region by using exploratory and descriptive method. To analyse the impacts on climatic conditions and its consequences on ecosystem services and livelihood, meteorological data of Ravi basin was used and chapter also reports the viewpoint of the respondents belonging to different age groups.

Dr. Ranjit Dutta and Dr. Sarada Kanta Sarma in their chapter, "The Environment Impact Assessment (EIA) Report of the 2000 MW Lower Subansiri Hydroelectric Project: A Critical Analysis" describe the Environment Impact Assessment (EIA) report of the 2000 MW Lower Subansiri Hydroelectric Project in North-east India prepared by the WAPCOS (Water and Power Consultancy Services) which establishes that downstream survey was done only up to 7 km from the dam site without giving much importance to the actual scenario and avoiding some most crucial ecological aspects. The EIA report gives some insufficient records on terrestrial flora, phytoplanktons and fish diversity. Especially the wetlands, associated and influenced by the unregulated Subansiri river has not found any place in the EIA report.

Interestingly, no mention of the Ganges Dolphin Platanista Gangetica Roxb. is found in the report, whereas Subansiri river provides a healthy habitat for a good number of these critically endangered fresh water dolphins. From the pre-impact study, it becomes clear that rich downstream ecology of the river with its present existing environmental scenario will be adversely affected due to the construction and operation of the hydel project and there will be distinct possibilities of elimination of some native flora and fauna. In addition, people's livelihood will also be affected largely through alteration of the flow regime of the river. Therefore in-depth study with comprehensive documentation of all biotic and abiotic parameters is obligatory before taking any decision about the operation of the 2000 MW Lower Subansiri Hydroelectric Project.

Dr. K.N. Bhatt in his chapter, "Climate Change and Disasters: Uttar Pradesh in Context" mentions that climate change and

Disaster Risk Reduction (DRR) are closely linked as the methods and tools of DRR provide powerful capacities for adaptation to climate change. The increased intensity of natural hazards as a result of climate changes turn into disasters when a combination of a vulnerable, exposed, and ill prepared population encounters a hazard event. This chapter analyzes the outcomes of Disaster Risk Management Projects (DRMP) implemented in Uttar Pradesh. It documents some case studies from project covered districts to learn lessons for policy interventions in the future. It also examines interlinkages between the incoming dangers of climate change and disasters. The Government of India (GoI)-UNDP initiated DRMP 2002-07. It dovetailed all national programmes, especially pro-poor initiatives, uncertainty reduction, and vulnerability adaptation and assessment covered under the United Nations Framework Convention on Climate Change (UNFCCC). A sub-component of DRMP focused on sustainable reduction in earthquake risk prone urban areas across the country. Both these projects have been implemented in Uttar Pradesh during 2004-2008. The study in all documented 31 case studies on risk management of floods, earthquakes, other disasters: practically covering all components of the DRMP.

Overall the projects have been successful in developing a community based organizational structure for tackling disasters which cause heavy damage in U.P. It successfully built an organizational structure in association with the NGOs and the community groups which may be crucial to maintain the sustainability of the DRMP in coming years. Mobilization of skilled human resources and volunteers has also become easier in the state with a strong network of various kinds of partners and stakeholders. Based on the study findings, some suggestions have been formulated for policy interventions under DRMP at the end of the chapter.

Dr. Sabu. M. Simon and Dr. Diraviaraj Maria Susai in their chapter, "Alternate Building Materials—An effective method to curtail Carbon emissions in Indian Scenario" highlight that climate change is a major global concern in the recent years wherein several studies are being carried out throughout the world to assess the various factors involved in climate change and the different means and ways to mitigate the same. It is well known fact that the huge concrete buildings are rated as one of the major reason for the ecosystem damage, greenhouse gas emissions and subsequent climate change. But not much of studies were carried out on the mitigation potential of alternate technologies of construction. If appropriate materials are used in the construction and renovation projects with a focus on reduced energy consumption and carbon dioxide emissions, the amount of energy saved would be enormous.

It highlights that in India, burnt bricks are the major building material used for construction and considerable quantum of carbon is emitted during preparation, baking and use of the bricks but points out that use of Stabilized Mud Blocks for construction is an alternate technology which is promoted by many of the NGOs for eco-friendly and low cost construction of houses. The soil type and processing of the soil differ considerably between the two. The chapter is an outcome of a detailed study which was conducted in different brick kilns and stabilised mud block making units in Dindigul, Tamil Nadu on their energy consumption, carbon emissions and the ecological footprints while making the two products. A comparative study was made on the usage of these two different building materials and the efficiency of stabilised mud blocks in curtailing carbon emission was assessed. The study in the manufacturing units forms the basis of this chapter.

Dr. Manish Mishra in his chapter, "Climate Change and its Impact on Local Indigenous Tribal Communities of Central India: A Case Study of Baiga Tribes of Madhya Pradesh" highlights that forests are the frontline in mitigating climate impacts by reducing exposures to scourging heat, dust storms and floods. Fluctuating weather conditions in India suggest that the country is reeling under climatic chaos. For more than a decade now, the country has been experiencing contrasting extreme weather conditions: from heat waves to cyclones, from droughts to floods. Orissa, Bihar for instance are states of a classic example of climate change events. It describes that several indigenous communities make their living within vulnerable environments—in mountainous areas, in forests or in dry areas and often the first to discern and suffer the effects of climate change. These people are not just victims of global warming; they have a critical role to play in supporting global adaptation to climate change.

Tribes of Dindori district acquire a distinctive knowledge of plant genetic diversity needed to fight plant and animal diseases and they know how to breed varieties that can cope with stressed environments. Local communities have always aimed to adopt variations in their climate. The tribal people living in and around world famous Kanha National Park possess knowledge about wild varieties of Agricultural crops (particularly Rice and Wheat) and other edible food crops. They are still growing various Agri-crops using organic manure since time immemorial. They are still growing Kodo and Kutki (rice) varieties besides its very low yield. They are utilizing wild tubers (yams) efficiently in the harsh conditions. This chapter thus discusses the local indigenous knowledge to cope up adverse climatic conditions.

We are confident that the chapters will add to the ongoing debate about the ill effects of global warming, its impact on the environment, changing lifestyles of human beings and importance of sustainable consumption patterns.

NOTES

1. Climate change is happening, http://www.epa.gov/climatechange/basics/
2. http://www.iso.org/iso/wsd2009/wsd2009_message.htm
3. http://unfcc.in/kyoto_protocol/

1

Climate Change

A Global Concern

Archana Wankhede

Introduction

Climate is the average pattern of weather for a particular region over a long period of time. Climate change refers to a significant variation in either the mean state of the climate or in its variability, persisting for an extended period, typically, decades or longer. Burning of coal, oil, and natural gas, deforestation, and various agricultural and industrial practices are altering the composition of the atmosphere and contributing to it. These human activities have led to increased atmospheric concentration of a number of greenhouse gases, including carbon dioxide, methane, nitrous oxide, chlorofluorocarbons, and tropospheric ozone in the lower part of the atmosphere. India is highly vulnerable to climate change as its economy depends mainly on the climate-sensitive sectors like agriculture and forestry. Its low-lying densely populated coastal areas are threatened by a potential rise in the sea level.

There are instances of people talking about how the weather is very different today than it was ten years ago. The earth's climate has changed many times over millions of years, sometimes slowly, sometimes very quickly. What makes climate change different today? From the very first coal-powered factories, people have known about the environmental and health effects of pollution. What we have only begun to realize in the past few decades is that our weather patterns are changing very quickly as a result of greenhouse gases. Greenhouse gases are produced by the burning of resources to create power and electricity for industry and residential needs. They trap sunlight and heat in the earth's atmosphere. Climate change can also happen on a smaller scale. Dense cities and suburbs are referred to as "heat islands", because the temperature in the city is noticeably higher than it is in the area around it. The plants, trees and grass help to soak up the heat of the sun and cool the air around them, but urbanization is quickly taking over these natural spaces. Instead, buildings, roads and parking lots absorb the sun's heat and reflect it back into the air.

Global warming does not mean that our planet will turn into a tropical island. Instead, it means that the weather we are used to, will become very unpredictable. Some places may experience droughts, while others will be flooded. Some places will become unbearably hot, while others will be covered by blizzards. As the ice caps melt, the level of the world's oceans will rise, permanently flooding many of the coastal areas where now people are living. As water temperature rises, there is a fear that the Atlantic Gulf Stream may fail, making most of Europe much colder than it is today. Even a small change in the temperature can have a powerful effect on species of animals that are not as adaptable as the humans. Many species are moving out of their traditional habitats as temperatures change. In Canada, grizzly bears are now found much farther north than they were in the past. This might not be a problem for the grizzlies, but it is a very big problem for the polar bears that already live there. This is happening to animals all-over the world, and as climate change becomes more dramatic it will also happen to humans. To prepare for this, and hopefully prevent it, there are many different international efforts to slow down the impact of climate change. The most famous, is the Kyoto Protocol. According to which countries that produce a lot of emissions will have "caps" placed on them, limiting how much greenhouse gases they can produce. If the country produces more emissions than they are allowed, they can buy "credits" from other countries that do not produce as much greenhouse gases. It is hoped that less developed countries will be encouraged to develop sustainably to earn credits that they can sell. Kyoto does not cover all greenhouse emissions, but it is still an important global step towards solving this important global issue. The existing business practices are preventing the Kyoto agreement from being implemented in many parts of the world, most noticeably in the United States, but many other countries are beginning to do their part in reducing their greenhouse emissions.

In developing countries like India, climate change could spell an additional stress on the ecological and socio-economic systems which are already facing tremendous pressures due to rapid urbanization, industrialization and economic development. The climate change studies already undertaken reveal that everyone in the community should be informed about climate change and suitable steps should be taken to prevent long term damage to India's water cycle. This is essential since livelihood of a vast population of India depends on agriculture, forestry, wetlands and fisheries as land use in these areas is strongly influenced by water-based ecosystems that depend on the monsoon rains. Any change to the water cycle may also cause an increase in the water-borne diseases. As agriculture and allied activities constitute the single largest component of India's economy, contributing nearly 27 per cent of the GDP. The climatic changes that affect rainfall pattern could have a profound impact on the production and fertility as 62 per cent of the cropped area is still dependent on rainfall patterns in a particular area.

The Earth's surface has warmed by more than 0.8 °C over the past century, and by approximately 0.6 °C in the past three decades. This seems to be a clear and convincing evidence for the anthropogenic climate change. This warming has been linked to more extreme weather conditions such as intense floods and droughts, heavier and more frequent storms, and a possible increase in the frequency and intensity of the El Niño Southern Oscillation. These changes are largely caused by human activities, mainly by the burning of fossil fuels, releasing carbon dioxide (CO_2) that traps heat within the atmosphere. These CO_2 emissions continue to rise, and climate models project that the average surface temperature will rise by 1.1°C to 6.4 °C over the 21st century.

Since 1990, WHO has published a series of reports on climate change and has participated in the review processes such as the Intergovernmental Panel on Climate Change. These activities have outlined four key characteristics of the health risks generated by warming and a more variable climate. First, these hazards are diverse, global and probably irreversible over human time scales. They range from increased risks of extreme weather, such as fatal heat waves, floods and storms, to less dramatic but potentially more serious effects on infectious disease dynamics, shifts to long-term drought conditions in many regions, melting of glaciers that supply freshwater to the large population centres, and rising sea levels leading to salination of sources of agriculture and the drinking water.

Secondly, the health impacts of the climate change are also potentially huge. Most of the important global killers are highly sensitive to climatic conditions. Malaria, diarrhoea and protein-energy malnutrition together cause more than 3 million deaths each year. Thirdly, these risks are inequitable, as the greenhouse gases that cause climate change originate mainly from developed countries, but the health risks are concentrated in the poorest nations, which have contributed least to the problem. Finally, many of the projected impacts on the health are avoidable, through a combination of public health interventions in the short-term, supported for adaptation measures in health-related sectors such as agriculture and water management, and a long-term strategy to reduce the human impacts on climate.

Similarly, the forests make a considerable contribution to the Indian economy contributing 1.7 per cent to the GDP of the country and non-timber forest products providing about 40 per cent of total forest revenues and 55 per cent of forest-based employment. Nearly 55 million people living in and around forests in India depend upon non-timber forest products as a critical component for their sustenance. Hence, common man should be communicated about how to protect and sustain environment to maintain a check on the climate change.

Awareness among Masses

The best way to make people aware about climate change is by giving:

- High priority and high interest by media on related issues.
- High focus and long-term vision of government agencies in protecting and promoting green technologies.
- Improving common man and science community communication.
- Strong political will to support clean environment issues.
- Translation and communication of these issues through multimedia in all regional languages in India.

- Highlighting good practices, glorifying achievements of people/institutions in preserving climate on Earth.
- Taking immediate steps to slow down climate change for fast socio-economic development of the people on sustainable basis.
- Ensuring that whatever we do, as populations expand and lifestyles change, the good condition are to be maintained in order to make available the same natural resources that we have to the coming generations. Natural resources include: fresh air, clean water, farmland, wildlife, forests, unpolluted seas and a stable climate.
- Roads, factories and housing should not destroy habitats of animals and plants.
- Bio-fuel development supported by new technologies need to be accelerated, aided by policies that provide direct incentives to invest in the most efficient bio-fuels.
- If we do not act now, we will leave a much larger problem to future generations. If all join hands to fight for this cause, we can reduce its impact on our planet Earth.

Further oil and natural gas take us down the street and around the world. They warm and cool our homes and businesses. They provide the ingredients for medicines, fertilizers, fabrics, plastics and other products that make life safer, easier and better. While we rely on them for most of our energy and will likely do so for years to come, emissions from their production and use may be helping to warm our planet by enhancing the natural greenhouse effect of the atmosphere. That is why oil and gas companies are also working to reduce their greenhouse emissions.

To start, they are closely managing their own energy needs use. One strategy involves heat and power technology that turns waste heat into energy, reducing energy consumption and emissions. The most recent data from the U.S. Energy Information Administration shows that CO_2 greenhouse emissions from U.S. industry, including oil and natural gas companies, have declined and were actually below 1990 levels. In 2006, for example, improvements in the energy efficiency at API member refineries—compared to the technology used in 2002—produced energy savings equivalent to taking more than 5,28,000 cars off the road, or savings equivalent to the electricity used by more than 9,50,000 homes.

The companies are also expanding use of alternative energy with lower greenhouse emissions. They are researching, developing and/or marketing virtually every form of renewable or alternative energy, including solar power, bio-fuels, geothermal energy, and wind power. They are working with the automakers and government agencies on new fuel/vehicle technology such as fuel cells and hydrogen power. While companies must continue to meet the demands of today's and tomorrow's consumers for oil and gas, they are also preparing for a future in which alternatives will play a much more significant role.

On other fronts, companies are reducing natural gas flaring to cut emissions (while also adding to energy supplies) and storing CO_2 underground, where it can be safely held for thousands of years. The oil and gas industry has also been implementing new emission estimation and tracking tools to enable it to assess how well it is meeting the goals it has set for itself and report progress to the public. And just as our industry is taking action, everyone can contribute to improving the energy efficiency. For instance, the "Power Is in Your Hands" campaign, created by 23 companies, trade associations and non-profits, provides consumers with information on what they can do to help. Its website offers valuable energy efficiency information for homes, transportation, and businesses. The campaign also encourages all Americans to take the Six Degrees of Energy Efficiency Web Challenge.

The oil and natural gas industry shares a keen interest in the policy issues arena. As demand for energy to keep our homes, vehicles, and businesses running continues to increase, so does our advancement in technology, allowing us to provide safe, reliable, and affordable energy. While serious challenges face our nation on a variety of fronts, oil and natural gas industry representatives remain actively engaged with government leaders to ensure informed decision-making so that the energy needs of tomorrow are met easily.

Greenhouse Gas Abatement: A Global Public Good

Public goods are differentiated from private goods by two characteristics. First, they exhibit the property of being non-exclusive. That is, it is difficult or impossible to prevent individuals from consuming the public good. A second property of pubic goods is that they are non-rivals. Therefore, consumption of the public good by one person does not reduce the amount of utility that a second person can obtain from the public good. The provision of public goods requires the collective action of all those individuals in society who demand the public good. However, because public goods are non-exclusive and non-rival, competitive markets are unable to provide socially optimal levels of public good provision. Thus, provision of public goods requires a mechanism for ensuring that each individual in society pay his or her fair share for providing the public good.

The global atmosphere provides a diversity of services to the world's populations that have economic value. One service is to regulate climate. The climate regulation function is both non-rival and non-exclusive and therefore can be characterized as a pure public good. Another service of the atmosphere is to assimilate the secondary outputs of economic development. Since there are limits on the ability of the atmosphere to assimilate wastes without affecting other atmospheric services (such as climate regulation), the assimilative services of the atmosphere are rival and subject to congestion. In the case of Green House Gas (GHG) emissions, congestion is not measured in terms of reduced capacity to assimilate more GHG but in terms of reductions in the climate regulation (i.e., reductions in a public good). Thus, increased concentrations of GHG into the atmosphere have externality characteristics. Assuming that reduced emissions lead to a more stable climate, and that the net benefits of a more stable climate are positive, then the abatement of greenhouse gas emissions prevents the reduction of climate regulation capacity, which can be considered to be a pure public good. In this context, therefore, reduced emissions can be viewed as externality correcting actions that lead to the creation of a pure public good.

There is no international agency with the power to assign responsibilities for emission limitation to individual nations and, in effect, require payment for the public good. Thus, the problem must be addressed by trying to find a solution through the cooperative actions of nations as has been done through the UNFCCC (United Nations Framework Convention on Climate Change) process. The Kyoto Protocol attempts to facilitate a strengthened collective response to deal with global warming through a long-term process of negotiations on targets involving a growing number of countries, including developing countries. It is clear that current commitments only address the issue of climate change in a limited way, and the success of the long-term vision of the Protocol will only be known in time. Thus, for the foreseeable future it is likely that there will be under provision of climate regulation services. There may be a number of reasons for this, but among them is the public good nature of the issue. The economic theory suggests that non-cooperation may be the dominant strategy of nations relative to the issues such as global warming that involve a pure public good. The result will be under provision of the public good.

Climate Change Solutions

There are number of ways by which we can put a check on climate change. After decades of denial, most of us now realize we are addicted to the fossil fuels that create vast quantities of carbon dioxide, and that this addiction also extends to a wide variety of industrial and agricultural practices that create a whole host of other global warming gases. We are fundamentally changing the planet we live on. Challenges are numerous but they can be treated with appropriate solutions. These are:

Fossil Fuels: The first challenge is eliminating the burning of coal, oil and, eventually, natural gas. This is perhaps the most daunting challenge as denizens of richer nations literally eat, wear, work, play and even sleep on the products made from such fossilized sunshine. And citizens of developing nations want and arguably deserve the same comforts, which are largely due to the energy stored in such fuels.

Oil is the lubricant of the global economy, hidden inside such ubiquitous items as plastic and corn, and fundamental to the transportation of both consumers and goods. Coal is the substrate, supplying roughly half of the electricity used worldwide—a percentage that is likely to grow, according to the International Energy Agency. There are no perfect solutions for reducing dependence on fossil fuels (for example, carbon neutral biofuels can drive up the price of food and lead to forest destruction, and while nuclear power does not emit greenhouse gases, it does produce radioactive waste), but every bit counts. Employing alternatives when possible as using plant-derived plastics, biodiesel, wind power and to invest in the change, be it by divesting from oil stocks or investing in companies practicing carbon capture and storage is need of the hour in the present scenario.

Infrastructure Upgrade: Buildings worldwide contribute around one third of all greenhouse gas emissions (43% in the U.S. alone), even though investing in thicker insulation and other cost-effective, temperature-regulating steps can save money in the long run. Electric grids are at capacity or overloaded, but power demands continue to rise. And bad roads can lower the fuel economy of even the most efficient vehicle. Thus, investing in new infrastructure, or radically upgrading the existing highways and transmission lines, would help cut greenhouse gas emissions and drive economic growth in the developing countries.

It is quite pertinent that it takes a lot of cement, a major source of greenhouse gas emissions, to construct new buildings and roads. The U.S. alone contributed 50.7 million metric tons of carbon dioxide to the atmosphere in 2005 from cement production, which requires heating limestone and other ingredients to 1,450 degrees Celsius (2,642 degrees Fahrenheit). Mining copper and other elements needed for electrical wiring and transmission also causes globe-warming pollution.

Move Closer to Work: Transportation is the second leading source of greenhouse gas emissions (burning a single gallon of gasoline produces 20 pounds of CO_2). One way to dramatically curtail transportation fuel needs is to move closer to work, use mass transit, or switch to walking, cycling or some other mode of transport that does not require anything other than human energy. There is also a option of working from home and telecommuting several days a week. Cutting down on long-distance travel would also help, most notably airplane flights, which are one of the fastest growing sources of greenhouse gas emissions and a source that arguably releases such emissions in the worst possible spot (higher in the atmosphere). Flights are also one of the few sources of globe-warming pollution for which there is not already a viable alternative: jets rely on kerosene, because it packs the most energy per pound, allowing them to travel far and fast, yet it takes roughly 10 gallons of oil to make one gallon of JetA fuel. Restricting flying to only critical, long-distance trips—in many parts of the world, trains can replace planes for short to medium distance trips which would help curb the airplane emissions.

Consume Less: The easiest way to cut back on greenhouse gas emissions is simply to buy less stuff. Whether by forgoing an automobile or employing a reusable grocery sack, cutting back on consumption results in fewer fossil fuels being burned to extract, produce and ship products around the globe. The crux is think green when making purchases. For instance, if you are in the market for a new car, buy one that will last the longest and have the least impact on the environment. Thus, a used vehicle with a hybrid engine offers superior fuel efficiency over the long haul while saving the environmental impact of new car manufactured. Paradoxically, when purchasing essentials, such as groceries, buying in bulk can reduce the amount of packaging—plastic wrapping, cardboard boxes and other unnecessary materials. Sometimes buying more means consuming less.

Be Efficient: A potentially simpler and even bigger impact can be made by doing more with less. The citizens of many developed countries are profligate wasters of energy, whether by speeding in a gas-guzzling sport-utility vehicle or leaving the lights on when not in a room. Good driving and good car maintenance, such as making sure tiers are properly inflated can limit the amount of greenhouse gas emissions from a vehicle and, perhaps more importantly, lower the frequency of the payment at the pump.

Similarly, employing more efficient refrigerators, air conditioners and other appliances, such as those rated highly under the U.S. Environmental Protection Agency's Energy Star programme, can cut electric bills while something as simple as weatherproofing the windows of a home can reduce heating and cooling bills. Such efforts can also be usefully employed at work, whether that means installing more efficient turbines at the power plant or turning the lights off when you leave the office.

Stop Cutting Down Trees: Every year, 33 million acres of forests are cut down. Timber harvesting in the tropics alone contributes 1.5 billion metric tons of carbon to the atmosphere. Thus represents 20 per cent of human-made greenhouse gas emissions and a source that could be avoided relatively easily. Improved agricultural practices along with paper recycling and forest management, balancing the amount of wood taken out with the amount of new trees growing could quickly eliminate this significant chunk of emissions. When purchasing wooden products, such as furniture or flooring, buy used goods or, failing that, wood certified to have been sustainably harvested. The Amazon and other forests are not just the lungs of the Earth; they may also be humanity's best short-term hope for limiting the climate change.

Unplug: The citizens spend more money on electricity to power devices when off than when on. Televisions, stereo equipment, computers, battery chargers and a host of other gadgets and appliances consume more energy when seemingly switched off, so unplug them instead. Purchasing energy-efficient gadgets can also save both energy and money—and thus prevent more greenhouse gas emissions. To take but one example, efficient battery chargers could save more than one billion kilowatt-hours of electricity $100 million at today's electricity prices and thus prevent the release of more than one million metric tons of greenhouse gases. Swapping old incandescent light bulbs for more efficient replacements, such as compact fluorescents would save billions of kilowatt-hours. There are many and varied solutions to the climate change which have been proposed, including individual frugality, energy conservation measures, renewable energy, and carbon sequestration. Each solution contributes towards the cause which is controllable yet alarming.

Energy Conservation will show the earliest payback in terms of CO_2 reductions—in many cases an investment in energy conservation made this year will show CO_2 reductions this year, and every year thereafter. Because we have been living in a world of artificially cheap energy for decades, there are huge opportunities for energy conservation.

Renewable Energy including energy from wind, solar, wave, biofuels, etc., substitutes directly for fossil fuels and eliminates CO_2 emissions entirely. A small note of caution is needed—in a few cases, most notoriously certain biofuels, a large amount of energy input is required to create renewable energy, in some cases even exceeding the resulting energy output. Most renewable energy, however, is extremely efficient, and is poised to grow in abundance due to the rising costs of fossil fuels. In many places where governments have stepped in to help this process along, renewables are already playing an important role.

Sequestration or the long-term trapping of carbon dioxide before it enters the atmosphere, is an intermediate step along the way, but is not a solution in and of itself. Carbon dioxide can be sequestered as a gas by pumping it underground or into the ocean, or it can be sequestered by plants—however carbon sequestered by plants is, in most cases, quickly released to the atmosphere again. The global carbon budget of plants can be changed to sequester a greater amount of CO_2 from the atmosphere, but so far we have been doing the opposite—cutting and burning forests for instance which releases vast amounts of CO_2 into the atmosphere, and poor agriculture resulting in CO_2 being released from the soil. Some of these changes are reversible in the long run.

There are many ways that we can make a difference, such as driving less, insulating one's house better, changing the voting priorities, buying organic food, eating less meat, buying fuel efficient appliances and vehicles.

REFERENCES

business.outlookindia.com
edugreen.teri.res.in
envfor.nic.
wikipedia.org/wiki/Climate change
www.climatechallengeindia.org
www.global-greenhouse-warming.com
www.globalissues.org
www.greenfacts.org
www.oecd.org
www.worldviewofglobalwarming

2

Impact of Climate Change on Food Security

A Study on Water Resource Management in Cachar District, Assam, India

Mithra Dey and *Nikhil Bhusan Dey*

Introduction

Climate change is one of the most serious problems the human civilization is confronting in the recent years. It is evident in the form of global warming, melting of glaciers, rise in sea level, unprecedented floods, drought or even cloud burst, increase in CO_2 concentration, rise in incidence of pests and pathogens and incidence of diseases, etc. Food security on the other hand, as defined by the Food and Agriculture Organisation (FAO) is "a situation when all people, at all times, have physical and economic access to sufficient, safe and nutritious food that meets the dietary needs and food preferences for an active and healthy life." Food security can be ensured only if agriculture is able to produce sufficient food to meet the demand and also provide income to enable the people to buy the required quantity of food. Agriculture is therefore not only an avenue to food but also a source of income.

Assam, one of the eight north-eastern states of India is extremely rich in natural resources and has a variety of flora and fauna. It is part of the Indo Burma Global Biodiversity hotspot. Economy of Assam is based mainly on agriculture and its allied activities. Assam falls in the **rainfed** area and agriculture is largely dependent on rainfall and natural water resources. The agricultural practices in the North-east India and also in Assam is broadly of two types: (1) Settled cultivation is practiced in the plains, valleys, foothills and terraced hill slopes, and (2) Shifting cultivation is done in hilly areas of the region. Agriculture includes mainly food crops of which rice is the dominant crop cultivated. Agriculture is mainly of subsistence level, cropping intensity is low mainly because of inadequate irrigation facilities and farm holdings are small.

The soil, topography, rainfall and climatic conditions of the State is suitable for paddy cultivation which occupies 89.47 per cent of the Net cropped area and 62.31 per cent of Gross cropped area. The normal paddy area of the State is 24.82 lakh hectare (Economic Survey Assam, 2007-2008). The other crops cultivated are wheat, maize, pulses and oilseeds. Among horticultural crops banana, pineapple, orange, litchi, papaya, lemon, jackfruit and mango are produced. Arecanut, chilli, ginger, turmeric, and garlic are also grown. Among plantation crop, the State dominates in Tea cultivation in the hilly terrain. Assam is the largest tea producer as well as consumer of tea in India. Agriculture is the backbone of Cachar and proper growth of agricultural sector will ensure food security and improve the socio-economic condition of the people in the district. The district has sufficient land and water resource which needs to be utilized and managed scientifically for improving yield. Safety measures have to be designed for times of intense floods and drought like conditions, both of which are becoming common in recent years.

This chapter tries to analyse these facts and puts forward some suggestions to manage the water resources along with rainwater harvesting with traditional and modern methods to achieve a higher yield in paddy and increase fish production.

Study Site: Brief Profile of Cachar District

River Barak and River Brahmaputra are the two major rivers of Assam. The District of Cachar is located in the southern most part of Assam in Barak Valley and is one of the oldest districts. Cachar district alone accounts for 54.7 per cent of the total area of the Barak Valley. The shares of other two districts of Barak Valley (i.e. of Hailakandi and Karimganj) in the total area are 19.2 per cent and 26.1 per cent respectively. The district encompasses an area of 3776 sq.km. and has a population of 14.45 lakh, as per 2001 census report (NABARD, 2010-2011). The district has three Agricultural Subdivisions, namely, Silchar, Sonai and Lakhipur. Of the total population, 86 per cent are rural and 80 per cent of them earn their livelihood from agriculture and its allied activities. Fishery, both capture and culture, forms a major occupation and contributes largely towards the economy. Other activities include handloom, handicrafts—cane and bamboo, migrant workers to mines, stone quarries etc.

Cachar is bounded by Manipur in the East, Hailakandi District and a part of Bangladesh in the West, Mizoram in the South and N.C. Hill districts and Meghalaya in the Northern side. It is located between 24°22′ and 25°8′ North Latitude and 92°24″ and 93°15′ East Longitude. The District is very strategically located, as the supply and communication lines to the entire Mizoram, Tripura and Manipur passes through it. Cachar abounds in wetlands of various sizes (locally called beels and haors), formed in the flood plains of River Barak and its tributaries, low hills (Tilla land) strewn in between and low-lying plains land. There are numerous ponds and tanks of varying sizes, streams, temporary pools and marshes in this predominantly low-lying area. River Barak flows a distance of 192 kms—east to west, receiving a number of tributaries *viz.*, Jiri, Chiri, Badri, Madhura, Jatinga, Hareng and Kalaincherra on the north and the Sonai, Rukni, Ghagra and Katakhal on the south. The river Barak itself bifurcates into the Surma flowing northwards and the Kushiara flowing southwards at Haritikar near Badarpur. Longai and Singla are the important tributaries of Kushiara while Ghumra is the most important tributary of Surma (Fig. 2.1.). All these rivers, especially the Barak which adopts a meandering path through the alluvial plains, frequently shift their courses and form ox-bow lakes (beels). River Barak also causes devastating floods nearly every year. The natural water bodies help in storing the excess water after heavy rainfall.

The District has a subtropical climate. The warm and humid climate generally prevails during the summer (June-September). During rainy season, the air is surcharged with moisture and rainfall is heavy. The annual rainfall of the district is around 2500 mm. The humidity ranges from 65-70 per cent during winter and 85-90 per cent during rainy season. The rainy season starts from May and continues up to October. Winter is short and mild from December to February. The months of March and April witnesses severe storms with strong wind, thunder and hail (locally called *Kal Baisakhi*).

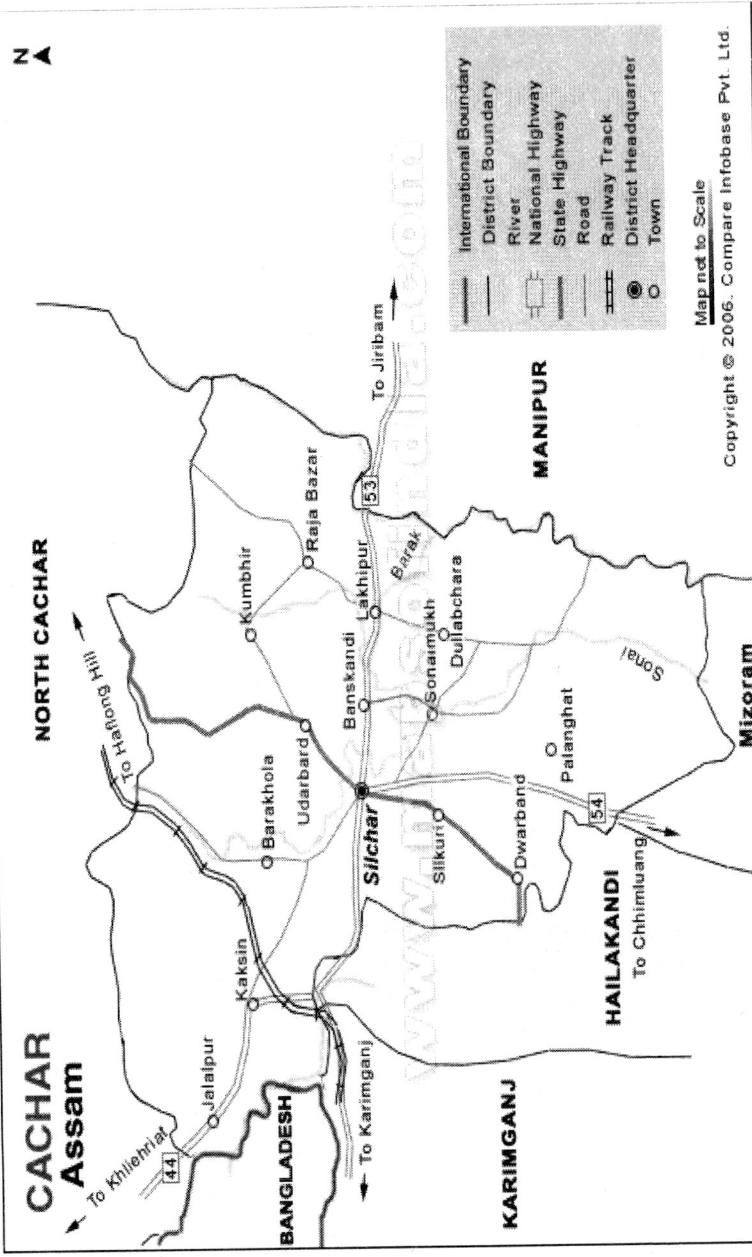

Fig. 2.1: Map of Cachar District, Assam.

Rainfall pattern in general reveals that 61 per cent of the rainfall is between June to September-Monsoon; 30 per cent is between March to May, i.e. Pre-monsoon and 9 per cent is between October to February i.e. Post-monsoon.

The soil of the district varies from alluvial to laterite soil. Texture is generally clay loamy to clay. The pH is acidic ranging from 4.5 to 6. The riverine tracts are found to be loamy to sandy loam soil. These soils are less acidic in reaction and pH ranges from 5.00 to 6.00.

Material and Methods

Data is mainly secondary, collected from various reports of Government Departments including NABARD. Data on rainfall, temperature and humidity was collected from Tea Research Association, Silcoorie Grant, Cachar.

Results

Agricultural Profile of Cachar

In the district of Cachar, agriculture is dominated by peasant subsistence and small scale mixed family farms. The net area sown and gross cropped area in the district are 1.25 lakh ha and 1.59 lakh ha respectively (NABARD, 2009-10) and the cropping intensity is 127 per cent. The average land holding is 0.25 ha. Irrigated land is only 0.1 per cent. Food crops dominate the entire cropping pattern in the district with paddy constituting 69.7 per cent of the gross cropped area and Tea being the major plantation crop in the district. The major food crop-paddy occupies 8,350 ha under Autumn (Ahu) paddy, 93,610 ha under winter (Sali) and 9,560 ha under Summer (Boro) paddy (Directorate of Economics and Statistics, Government of Assam, 2009).

Use of HYV seeds is less with only 40 per cent of the total area under paddy cultivation using HYV seeds. In the district, three varieties of rice are grown *viz.* Autumn Rice (Ahu), Winter Rice (Sali) and Summer Rice (Boro). Ahu is sown in the month of March-April and transplanted in April-May, it is harvested in September. Sali is transplanted in July-August and harvested in Nov.-Dec. Boro paddy or summer paddy is grown in low-lying land during dry season i.e., Nov-Dec. and harvested in April-May. Still another variety of winter rice (locally called Birain) is cultivated in low lying areas. Average rice yield (in Kg/ha) for 2004-05 to 2008-09 is presented in Table 2.1. Sali or Winter rice occupies 82 per cent of land under paddy cultivation. The pattern of cropping is mainly mono cropping and fields lie fallow in most cases after harvesting of paddy. Some farmers go for cultivating of pulses, oilseeds and vegetables. During winter season most of the farmers cultivate vegetables in the riverine tract.

Table 2.1: Area under Cultivation (in ha), Av. rice yield (Kg/ha), Production (MT) during 2004-05 to 2008-09 in Cachar District

Year	*Type*	*Area under cultivation (in ha)*	*Av. rice yield (Kg/ha)*	*Production (MT)*
2004-2005	Autumn (Ahu)	10174	214	2140
	Winter (Sali)	82968	1817	148467
	Summer (Boro)	9990	1485	14835
2005-2006	Autumn (Ahu)	9411	1961	18157
	Winter (Sali)	90194	2070	183913
	Summer (Boro)	10261	1630	16722
2006-2007	Autumn (Ahu)	4070	1584	6344
	Winter (Sali)	89689	1711	151116
	Summer (Boro)	7819	1036	8100
2007-2008	Autumn (Ahu)	8427	2100	17697
	Winter (Sali)	89544	2200	196997
	Summer (Boro)	8759	1800	86670
2008-2009	Autumn (Ahu)	8350	1681	13808
	Winter (Sali)	93610	2166	199711
	Summer (Boro)	9560	2210	21126

Source: Directorate of Economics and Statistics, Govt. of Assam, 2009-2010.

The major constraints in the rice cultivation are:

The crops are damaged when flood water inundates the paddy fields. Floods are frequent and autumn crop is seen to suffer most. During the current year, there has been excess rainfall and fields are waterlogged which has made it difficult for the farmers to sow Sali paddy. According to a report published in *Dainik Jugosankha*, a local daily newspaper (Das, 2010) nearly 40 per cent of land has remained unused due to unprecedented rainfall and waterlogging. It is also expected to have adverse effect on cultivation of other crops like pulses, vegetables, etc. during the winter season. There exists uncertainty in the production also when there is low rainfall-like in the year 2004 and 2006 when there was drought like condition in Cachar (Table 2.2). Lack of irrigation facilities prevents full utilization of the available land that can be brought under cultivation, only about 0.1 per cent of land receives irrigation.

Due to lack of technical know-how and irrigational facilities farmers are reluctant to adopt HYV seeds, use more fertilizer and pesticide, which would require more investment.

As a result, agricultural production from all sources remains far behind the potential the district holds. Cultivation of autumn rice has also declined and many prefer to go for cultivation of Sali or winter rice.

Table 2.2: Paddy Production (MT) in Cachar District and Assam (1999-2000 to 2008-2009)

Year	*Cachar*	*Assam*
1999-2000	181850	3738000
2000-2001	221685	3894000
2001-2002	224744	3999000
2002-2003	220058	3861000
2003-2004	234639	3881000
2004-2005	165442	NA
2005-2006	218792	3552000
2006-2007	165560	2916000
2007-2008	301364	2550000
2008-2009	234645	NA

Source: Directorate of Economics and Statistics, Govt. of Assam, 2009-2010.

Fish Farming in Cachar

Fisheries is another important economic activity of Assam and also in Cachar district where fish is an integral part of the diet. Fish also has cultural importance. Cachar by virtue of its topography is bestowed with large number of natural water resources which can be utilized successfully for culturing fish. The annual requirement of fish in the district is estimated at 29,678 MT at 18-20 Kg/head/annum. The annual production of fish is around

18,000 MT out of all the resources in the district. Thus, there exists a large gap between the demand and supply of fish, which is fulfilled by importing fish from Andhra Pradesh, Uttar Pradesh, Bihar and West Bengal. Nearly 5000-6000 tonnes of fish is imported annually. At the State level, fishery sector contributes to 2.47 per cent of the Gross Domestic Production but still there remains a shortfall of nearly 10,678 MT which has to be fulfilled by importing fish from other states.

The natural water resources in the district can provide sufficient fish to meet the demand in the district (Table 2.3), but various constraints are responsible for the shortfall.

Table 2.3: Fisheries Resource in Cachar district

Sl. No.	*Resource*	*No. of Resources*	*Water area (Ha)*
1.	Beel Fisheries	135	3792.57
2.	Riverine Fisheries	29	2398.8
3.	Forest	91	245
4.	Tea garden	81	230
5.	Low lying areas	268	1400
6.	Paddy fields	691	968
7.	Ponds and tanks	20507	6159
8.	Community tanks	20	50
	Total Water Area		27843.37

Source: District Fishery Office, Cachar.

The total production is around 790 Kg/ha only. However, keeping in mind the total water potential available, the production can be significantly increased. Table 2.4 presents the fish production in Cachar between 2002-03 to 2008-09.

Table 2.4: Fish Production (in MT) in Cachar district between 2002-03 to 2008-09

Year	*Fish Production (MT)*
2002-03	11947
2003-04	13120
2004-05	17249
2005-06	18329
2006-07	17939
2007-08	16950
2008-09	18062

Source: District Fishery Office, Cachar.

The major constraints in fish farming are as follows:

1. Low technical know-how among the fish farmers and non-adoption of modern technology.
2. Inadequate supply of quality seed/fry and fingerlings. During the year 2008-09, fish seed production in the district was 47.5 million nos. against the demand of 60 million numbers. The number of ecohatcheries is 7, of which one is owned by the fishery department and the rest are privately owned. The number of seed produced is not adequate and also has to be fetched from a long distance.
3. Climatic conditions also need to be addressed properly. On one hand, there is recurrent problem of floods in the district which creates acute problem for the entrepreneurs who adopt fish culture. Fish farmers suffer huge loss during floods and hence become reluctant to make big investment. In the recent years, there is a declining trend in the rainfall in the region, such erratic rainfall and more number of rainless days create difficulties in production of fish.
4. Lack of finance and insurance coverage from companies to entrepreneurs prevent them from taking up fish farming as an economic activity.
5. Derelict condition of beels and haors due to siltation, erosion, and degradation leads to underutilization and production becomes less. Polluted water due to improper solid waste disposal is another serious problem for the wetlands.
6. Sale of immature fish due to poor economic condition of the fish farmers is yet another problem of fish farming in Cachar District of Assam.
7. Cachar District has plenty of hillocks which are used for tea cultivation, where pesticides and fertilizers are used and these enter into aquatic bodies causing huge damage to the fish population and may also affect human health.

Observations on Climatic Conditions in Cachar District

Rainfall in the whole of North-east India has always been adequate and sometimes more than the expected, creating floods and destroying human property and agriculture. The rainfall in NE India ranges between 2000 mm to 3000 mm (Das *et al.*, 2004); Mawsynram and Cherrapunji in Meghalaya have the highest rainfall in the world. Recently erratic rainfall is being observed both in terms of frequency and occurrence, making cropping a difficult task. There is a rise in monsoon rainfall (June-Sept) and a decrease in the winter rainfall (Dec-Feb). The data on rainfall in Cachar District shows peak rainfall of about 439 mm-687 mm in monsoon season and around 0 mm rainfall in December. There is a declining trend of annual rainfall in the Barak Valley Region (Fig. 2.2) with more number of rainless days and scorching heat. This is accompanied with fewer days of severe rainfall leading to inundation of lowlying plains land damaging crops. Both the situation creates difficulty for the farmers. Paddy production in the district was low in 2004-2005 when the district had severe and successive floods leading to damage of paddy cultivation. Again in the year 2006, there was low rainfall and a drought like situation in Cachar along with several other districts of Assam which, along with poor irrigational facilities, hit paddy production. Sali or Winter paddy was seriously affected. Due to uncertainties autumn rice cultivation in the district has gone down with more farmers opting for winter rice cultivation. Since crop cultivation in the *Kharif* season is faced with natural risks such as floods, the use of

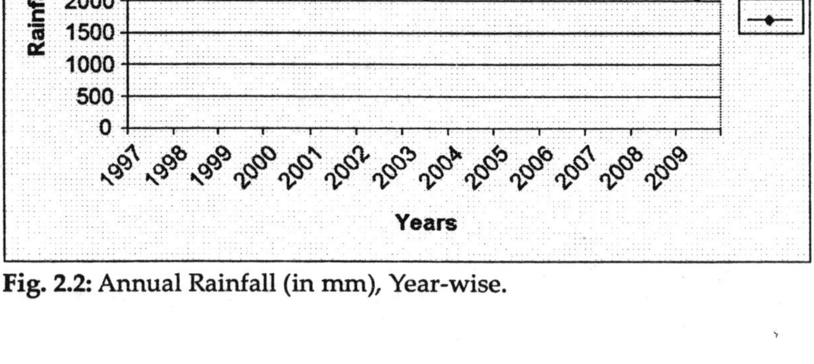

Fig. 2.2: Annual Rainfall (in mm), Year-wise.

HYV seeds and chemical fertilizers and pesticides has been low, resulting in lower yields and paddy cultivation remains in the subsistence level in spite of opportunities available. Therefore, *Rabi* season cultivation needs to be promoted with a continued emphasis on irrigation through shallow tube wells. Rainwater can be stored in the aquatic bodies and used for irrigation. Watersheds can be developed. Cultivation of vegetables, fruits, and spices and commercialization of agriculture needs to be encouraged by investing in marketing and storage facilities.

Assam can produce enough agricultural surplus through appropriate technological intervention designed for increasing cropping intensity from its present level of about 146 per cent to at least 200 per cent (state level), and evolving a cropping pattern consistent with its agro-climatic factors. This, however, requires assured irrigation. Assam is enormously rich in groundwater resource on the basis of which, another 47 per cent of the net sown area of the state can be brought under assured irrigation. Considering the fact that the district has adequate surface water resource, irrigation can be done using the surface water.

According to reports (Das, *et al.*, 2004) rainfall in the region was scantiest in 2006 (considering the data for a period of last 25 years) since 1982. Again worst situation prevailed in 2009 when there was deficit rainfall affecting transplantation and sowing of paddy. According to South Assam Meteorological Subdivision, a significant change of seasonal rainfall has been observed in the region. The summer rainfall is found to be decreasing over the region significantly during the last century, approximately 11 mm per decade (Mirza, *et al.*, 1998). The temperature and per cent relative humidity for 1997-2009 (Fig. 2.3 and Fig. 2.4) are both significantly high.

Temperature in summer ranges between 30 °C-34 °C (Max) and 24 °C-26°C (min) and during the winter months temperature ranges between 10 °C-13 °C (min) and 20 °C-25 °C (max). Per cent RH remains very high reaching up to 98-99 per cent in monsoon season. High temperature along with high humidity increases the possibility of pest attack and damage to the crops. Again both factors make it difficult for farmers to work in the fields. Das (2004) have observed a rise of +0.11 °C per decade and an annual mean temperature rise of 0.04 °C per decade in their study on climatic changes in the North-eastern region. During the current year (2010)

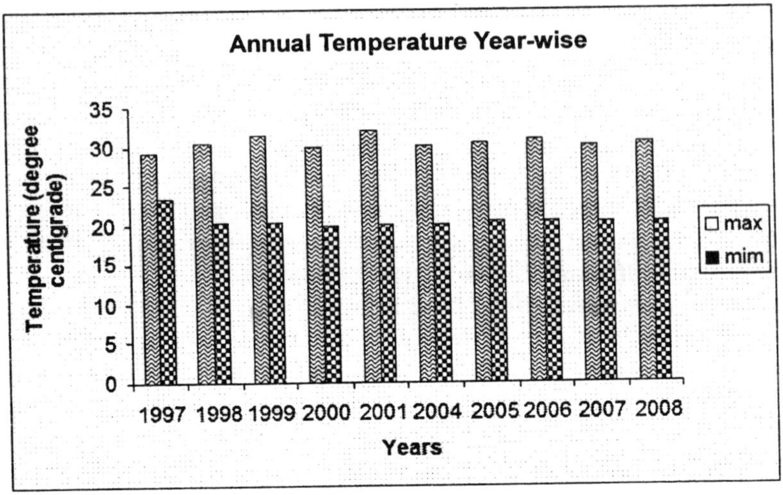

Fig. 2.3: Annual Temperature (°C), Year-wise.

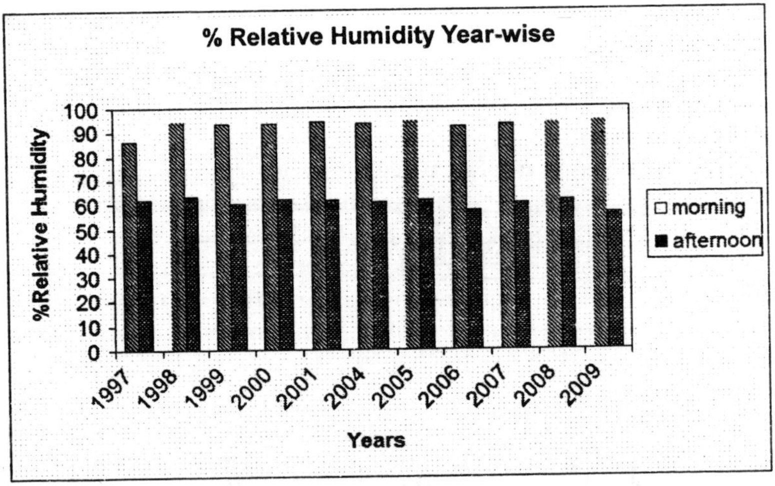

Fig. 2.4: Per cent Relative Humidity, Year-wise.

heavy rainfall and waterlogging of fields has delayed sowing and transplantation of Sali paddy in the District.

Some Suggestions

Under the present scenario of changing climatic conditions, it is essential to understand how to manage the situation so as to ensure sufficient agricultural production and food security for the people. It is necessary to formulate strategies which will help improve the agriculture and allied activities in the District. The following suggestions are put forward:

1. An important factor is that in spite of deficit rainfall, the incidence of floods remain a threat; deforestation and soil erosion cause siltation of low lying-land and natural water bodies which otherwise can store excess rainwater. The available natural water resources need to be properly managed for storage of excess water and can provide water for irrigation.
2. Flood mitigation has to be improved and given serious thought. The river bed has risen and that makes inundation a regular feature. Rainwater harvesting using traditional methods of water harvesting has to be revived for efficient water management.
3. Rainwater harvesting should be taken up to fulfil water demands; except Mizoram other states of the region do not utilize rainwater and the water is wasted. Watershed development projects need to be undertaken with government initiatives. Irrigational facilities are virtually non-existent in the District but can be developed with proper initiatives of the stakeholders i.e., agriculturists and fish farmers.
4. Derelict water bodies must be restored which will help increase in fish production. Wetlands are a habitat for variety of aquatic flora and fauna which too will be restored and provide sustainable livelihood to the rural population.
5. Ponds and ditches which are seasonally flooded can be converted into perennial ponds through deepening and expansion and can be used for fish culture.
6. Farmers should be encouraged to take up paddy-cum-fish culture so as to have dual benefit of paddy cultivation as well as produce fish with common investment. It can be introduced in low-lying areas along with Autumn and Sali rice.
7. The fisheries and agriculture department should together arrange for workshops and disseminate awareness

regarding modern technology to both the farmers and the fish farmers so as to ensure food availability to all.

8. Cooperatives, NGOs and SHGs can help the rural people in this regard. Community participation is essential, involving all stakeholders to manage water resources efficiently.

Conclusion

Climate change has emerged as one of the prominent global problems. Among various aspects, rainfall pattern is found to be undergoing changes and an attempt has been made to put forward some suggestions to improve agriculture and fish production in the Cachar District.

REFERENCES

Das, N. (2010). Baraker lukochuri te Cachar er 40% jami chash hin. *Jugasankho* (Daily Bengali newspaper), Sept. 9th, 2010, p. 5.

Das, P.J. (2004). Rainfall Regime of Northeast India: A Hydrometeorological Study with Special Emphasis on the Brahmaputra Basin. Unpublished Ph.D. Thesis. Gauhati University.

Das, P.J. (2009). Water and Climate Induced Vulnerability in Northeast India: Concerns for Environmental Security and Sustainability. WATCH Research Report 1. AARANYAK. Guwahati, Assam, India, www.aaranyak.org

Directorate of Economics and Statistics, Government of Assam. 2009.

Economic Survey Assam, Government of Assam. 2007-2008.

Mirza, M.M.Q.; Warrick, R.A.; Ericksen, N.J. and Kenny, G.J. (1998). "Trends and persistence in precipitation in the Ganges, Brahmaputra and Meghna river basins." *Hydrological Sciences-Journal-des Hydrologiques.* Vol. 43(6): 845-858.

NABARD, Potential Linked Credit Plan, 2009-10. District Cachar, Assam.

NABARD, Potential Linked Credit Plan, 2010-11, District Cachar, Assam.

www.mapsofindia.com

3

Role of Catalytic Converters in Controlling Pollution

Sanchita Chauhan

Introduction

One of the present environmental concerns is the ever increasing concentrations of pollutants released to the atmosphere from various sources. A health study report published in 1996 by the World Bank stated that in India due to air pollution 40,000 premature deaths and 17 million respiratory hospital admissions occurred in 1995. Such annual health incidences in Indian cities are caused due to ambient air pollution levels, exceeding World Health Organization guidelines. Hydrocarbons (HCs) form a significant part of these pollutants. Catalytic combustion of HCs is becoming an increasingly viable technology in power generation, domestic heating appliances and chemical process heaters (Dupont *et al.*, 2002). Monolithic catalytic converters are being extensively used to reduce such emissions from the vehicular exhaust.

Monolith catalysts have some common features in most of the applications they are used for, these include low pressure drop especially under high fluid throughputs, elimination of external mass transfer and internal diffusion limitations, low axial dispersion and high product selectivity, larger external surface, uniform distribution of flow (gas phase), elimination of fouling and plugging, and thus extended catalyst lifetime and easy scale-up (Chen *et al.*, 2008).

Types of Catalytic Converters

Different types of converters have been employed over the period of time, each type using a slightly different method and chemistry to reduce the harmful pollutants from the exhaust. Early model converters used a pelletized catalyst, but most modern converters are now designed with a free-flowing honeycomb ceramic catalyst. The type of converter required on a particular vehicle varies with model year, engine size and vehicle weight. Some of these converters are discussed below along with their functioning.

Two-Way Catalytic Converter (2WC)

Two-way catalytic converters were the first generation of catalytic converters introduced in 1975 (Gandhi *et al.*, 1982) for meeting the EPA emission standards for car exhausts. They were meant solely for the oxidation of unburnt Hydro Carbons and CO into H_2O and CO_2. The introduction of air through the pump in the exhaust manifold ensured that, no matter how the engine was running, enough air was added to maintain excess oxygen, required for efficient conversion (Shelef and McCabe, 2000). In them precious metals, platinum (Pt) and palladium (Pd) were used, supported on fine particles of high surface area alumina and were coated onto porous alumina pellets or inside monolith channels.

The major reactions taking place in these converters included:

$$CO + 0.5\ O_2 \rightarrow CO_2 \quad \text{... (1)}$$
$$C_xH_y + (x + (y/4))\ O_2 \rightarrow x\ CO_2 + (y/2)\ H_2O \quad \text{... (2)}$$

However the oxides of nitrogen (NO_x) passed through the converter unreacted.

Dual-Bed Catalytic Converter

Efforts to reduce NO_x emissions by lowering compression ratios or by exhaust gas re-circulation in the existing 2WCs could not be met with. It was observed that both CO and HC get converted under lean oxidizing conditions (i.e. with excess air) while NO_x reduction to N_2, required excess fuel (Shelef and McCabe, 2000). These opposing conditions led to further modifications of the converter.

The dual-bed converter introduced in 1978 (Cooper, 2002) consisted of two catalyst beds in series. The first bed was a reduction catalyst that catalysed the conversion of NO_x into N_2 as the vehicle was operated on the rich side of the stoichiometric point (air to fuel ratio maintained as 14.7:1). The most probably favoured reduction reaction taking place in the converter is shown below:

$$NO + CO \rightarrow 0.5\ N_2 + CO_2 \quad \ldots (3)$$

Additional air supply was provided just ahead of the second bed containing an oxidation catalyst to ensure complete oxidation of Hydro Carbons and Carbon Oxide as given by equations (1) and (2).

Despite both oxidising and reduction conditions being met by the dual-bed converters, this methodology had many limitations. In the reduction zone it was observed that rather than the formation of N_2 as desired during less oxygen conditions, NH_3 was being produced which was further oxidized back to NO_x in the oxidizing zone, limiting the effectiveness of the combined system (Shelef and McCabe, 2000). Also the fuel economy was adversely affected and engine had to be operated at a higher fuel/ air ratio. This constrained the operating range of the engine.

Three-Way Catalytic Converter (TWC or 3WC)

The shortcomings of the dual-bed converter were overcome with the introduction of TWC in 1981 (Cooper, 2002). In them rhodium (Rh) catalyst was introduced for reduction of NO_x along with Platinum and Palladium catalysts in a single converter assembly for bringing about simultaneous reduction of NO_x into N_2 and oxidation of unburned HCs and CO as shown by reactions (1), (2) and (3).

Efficiency of TWCs to conversion of these gases is primarily dependent on the mass ratio of air to fuel (λ) in the mixture leaving the exhaust manifold and entering the converter, the velocity of the exhaust mass and the temperature of the catalyst. In case of excess fuel (λ<1) i.e., at rich air fuel ratios, there is less conversion of CO and HCs, as they do not get fully oxidized due to less air. At conditions of air excess (λ>1) i.e., at lean air fuel ratios, there is an increase of the emissions of NO_x (Rijkeboer, 1991). By careful trimming of the catalyst composition, the reduction and oxidation conversions can be made to overlap at λ=1, so that complete treatment of the exhaust gas became possible within a narrow "lambda-window" (Rijkeboer, 1991). In case the air to fuel ratio is maintained at the desired stoichoimetric air fuel ratio of 14.7:1 then even a single bed containing the desired catalysts Pt, Pd and Rh could produce very high efficiency (up to 90 per cent) by simultaneous reduction of CO, HCs and NO_x.

Under steady, warmed-up conditions Three Way Catalytic perform very well. However, there are some circumstances, which may lead to undesirably high levels of emissions, these include the warm-up period before the catalyst is hot enough to work effectively, chemical or thermal deactivation of the catalyst and transient driving conditions. The effectiveness of a catalytic converter is also dependent on the geometric surface area (depends on the cell density), the thermal capacity (depends on the wall thickness), mass density and specific heat, washcoat and the type of catalyst used and the catalyst loading.

Open-Loop and Closed-Loop Systems

The TWCs could be further classified as open-loop and closed-loop converters. European solution of the open-loop three-way catalyst to reduce pollution consisted of the engine running rich (catalyst would reduce) during acceleration and the engine running lean (catalyst would oxidize) during cruise (Rijkeboer, 1991). This solution provided the emission balance between CO and HC on the one hand and NO_x on the other, as being very sensitive to the basic tuning of the engine.

Since the catalyst in an open-loop situation is either reducing or oxidizing at one time its performance is not as good as that of a catalyst in a closed-loop situation. In the closed loop system an exhaust gas oxygen sensor is used which senses whether the exhaust gas mixture is net oxidizing or net reducing at any given time and sends a signal back to the electronic engine control module. This module in turn signals the fuel injectors to increase or decrease the fueling rate as needed to drive the exhaust gas mixture back towards the stoichiometric point (Shelef and McCabe, 2000). Although the cost of this closed-loop control system was significant but automotive engines using closed-loop concepts are preferred as the closed-loop allows a fine tuning of operating conditions (Rijkeboer, 1991).

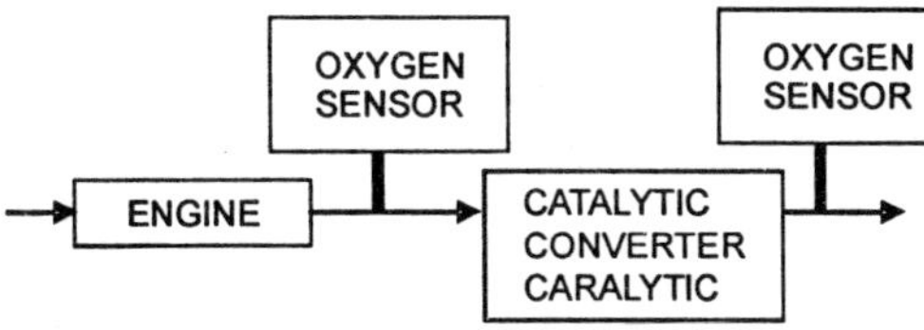

Fig. 3.1: Dual Oxygen Sensors for on-board Diagnostics (Farrauto and Heck, 1999).

Around 1994 On-Board Diagnostics II (OBDII) was introduced (Aimard *et al.*, 1996) to monitor the vehicles emission control components for increasing efficiency further by making use of two oxygen sensors as shown in Figure 3.1. One of the oxygen sensors mounted in the exhaust manifold, is used for feedback to the module for fuel control and the other oxygen sensor located at the downstream from the converter, gives an indication of the efficiency of the converter. The catalytic converter efficiency is measured by monitoring the oxygen storage capacity of the converter during closed loop operation. The downstream sensor is less prone to thermal aging or contamination and senses more fully equilibrated exhaust gas mixture than the upstream sensor (Shelef and McCabe, 2000).

Pre-Converter with Main Converter

The converter requires reaching a minimum operating temperature before effective conversion of the pollutant can take place. It is during this warm-up period from the cold start that the catalytic converter is least efficient and the vehicles expel most of the pollutants untreated. Some vehicles employ a pre-converter in the exhaust system immediately after the manifold to help reduce pollutants mainly Hydrocarbons during this warm-up period. So the primary function of this catalyst is to oxidize the Hydro-carbons during the first one-two minutes of the start of the converter.

The pre-converters proximity to the engine allows it to heat up quickly and start functioning faster (almost within 10s, Farrauto and Heck, 1999) as compared to the main converter. Further, it also pre-heats the exhaust gases going to the main converter which is a larger Three Way Converter located downstream in an under-floor position.

The pre-converter's small size (about 10 to 30% of the main converter volume) not only helps in faster heating but also allows for its easy installation close to exhaust manifold without any modification to car underfloor (Koltsakis and Stamatelos, 1997). The common formulations for pre-catalysts are Pd-Rh or Pd-only catalysts containing high precious metal loading (usually three times the loading of main converter) in the range 150 g/ft^3 (Koltsakis and Stamatelos, 1997).

The close couple design, however, exposes the catalyst to hotter overall temperatures, increasing the likelihood of carrier and metal sintering. Other disadvantages include space requirements and heat management in under hood location, aggressive driving modes of high speed and load, adding additional high temperature stress to close coupled catalysts (Farrauto and Heck, 1999).

Modeling of Catalytic Converter

Modeling catalytic converter helps in understanding the existing models by interpreting the reactivity data (obtained experimentally) to help in predicting the behaviour of monoliths in operation. In addition to this modeling also plays an important role in seeking improvements to be made in the already existing designs (Kolaczkowski, 1999). In order to reduce the design cost and time, modeling of three-way catalytic converter operation is used increasingly in the optimization of automobile converter systems. The optimization of the converter relies on catalytic material selection and chemical and mechanical engineering for its implementation.

The quality of the mathematical model is highly dependent on the assumptions made, numerical techniques employed and the validity of the physical and chemical parameters (kinetic expressions, physical properties, etc.) used in them (Kolaczkowski, 1999). So there are many challenges that the developer of a mathematical model of a catalytic monolith converter has to undergo. The challenges include the selection of appropriate kinetic expressions, recognition of the role that intraphase diffusion may play, pilot scale studies, the selection of heat and mass transfer correlations, and the modeling of transients (Kolaczkowski, 1999). These challenges are mostly based on areas that have not been addressed adequately by researchers over the last few years. For some of these areas solutions exist, while others require a long-term research. These challenges emphasized the need for the developers of mathematical models for converters to work in collaboration with the pilot scale experimentalists and researchers dealing with new catalyst systems, to predict accurate performance of converters.

A complete mathematical model of a monolith catalyst needs to include the coupled equations for transport of heat and mass for monolith channels and the porous walls between them. As in actual practice there are hundreds of reactions taking place in the converter, so modeling of the behaviour of such systems is a very difficult challenge as these complex competing reactions are strongly affected by both heat and mass transfer. So a number of areas that need to be addressed by scientists include providing adequate kinetics data covering a significant number of reactions and also data for the multitude of catalyst formulations and washcoats employed in automotive applications (Koltsakis *et al.*, 1997). Moreover, the situation is further complicated by the effect of various modes and degrees of catalyst ageing during vehicle operation. Models will continue to be refined over the period of time but their development should be guided by realistic description of the operating conditions. Presently, no available computer code can completely meet the objectives mentioned above and there is a need for model improvement and validation.

The following assumptions are usually made during modeling:

(1) Uniform monolith channel size and shape distribution for a single channel (along the flow direction) as well as for the whole reactor.
(2) Uniform gas distributions among the channels.
(3) Uniform distribution of catalyst activity.
(4) Heat transfer by radiation is negligible.
(5) No entrance effect on flow dynamics.

With these assumptions, only mass and heat balance equations need to be considered for a single channel (Chen *et al.*, 2008).

The exhaust comprising of Hydro Carbons, propylene (fast oxidizing HC) and propane (slow oxidizing HC) (Montreuil *et al.*, 1992) is considered. These gases are assumed to be present in the exhaust in the ratio C_3H_6/C_3H_8 : 1/2. In the converter these gases undergo the following highly exothermic oxidation reactions:

$$C_3H_6 + 4.5\ O_2 \rightarrow 3\ CO_2 + 3\ H_2O \quad \dots (4)$$

$$C_3H_8 + 5\ O_2 \rightarrow 3\ CO_2 + 4\ H_2O \quad \dots (5)$$

For both these gases heterogeneous phase reactions are considered. Platinum suspended in alumina washcoat is taken as the catalyst. The rate expressions used for propylene and propane oxidation reactions (Ahn *et al.*, 1986; Hayes *et al.*, 1992; Wanker *et al.*, 2000) are given in Table 3.1.

Table 3.1: Reaction Rates for Catalytic Reactions

Reaction Rate: $(-r_i)_{cat}(C_{s,i}, T_s) = k_{icat}\exp^{(-E_{icat}/RT_s)} C_i$ where i =1 represents propylene (C_3H_6) and i = 2 is propane (C_3H_8)	
K_{1cat} 9.14 X104 cm/s	K_{1cat} 50,242 J/g mole
K_{2cat} 2.40 X107 cm/s	K_{1cat} 89,849 J/g mole

One-dimensional Modeling for Two-gas System

At quasi steady state, the time derivative terms in gas phase mass and energy balance are not considered.

$$\text{i.e.,} \quad \frac{dC_{g,1}}{dt} = 0, \frac{dC_{g,2}}{dt} = 0, \text{ and } \frac{dT_g}{dt} = 0 \quad \dots (6)$$

Mass balance for gaseous phase:

For gas-1 $$v\left(\frac{dC_{g,1}}{dx}\right) + k_{g,1}S\left(C_{g,1} - C_{s,1}\right) = 0 \quad \dots (7)$$

For gas-2 $$v\left(\frac{dC_{g,2}}{dx}\right) + k_{g,2}S\left(C_{g,2} - C_{s,2}\right) = 0 \quad \dots (8)$$

Mass balance for solid phase:

For gas-1 $$a(-r_1)_{cat}\left(C_{s,1}, T_s\right) = k_{g,1}S\left(C_{g,1} - C_{s,1}\right) \quad \dots (9)$$

For gas-2 $$a(-r_2)_{cat}(C_{s,2},T_s) = k_{g,2}S(C_{g,2}-C_{s,2}) \quad \ldots (10)$$

Energy balance for gas phase:

$$-v\rho_g Cp_g\left(\frac{dT_g}{dx}\right) - hS(T_g - T_s) = 0 \quad \ldots (11)$$

Energy balance for solid phase:

$$\rho_s Cp_s\left(\frac{dT_s}{dt}\right) = a(-\Delta H_1)(-r_1)_{cat}(C_{s,1},T_s) + a(-\Delta H_2)(-r_2)_{cat}(C_{s,2},T_s) + hS(T_g - T_s) + \lambda_s\left(\frac{d^2T_s}{dx^2}\right) \quad \ldots (12)$$

Initial and boundary conditions:

$$C_{g,1}(0,t) = C^0_{g,1} \text{ and } C_{g,2}(0,t) = C^0_{g,2}$$

(entering gas concentrations) ...(13)

$$T_g(0,t) = T^0_g \text{ and } T_s(x,0) = T^0_s$$

(entering gas temperature and solid temperature at start) ... (14)

at $x = 0,\ \dfrac{dT_s}{dx} = 0$ (lagging solid catalyst at entry) ... (15)

at $x = L,\ \dfrac{dT_s}{dx} = 0$ (lagging solid catalyst at exit) ... (16)

The above equations form a set of coupled ordinary and partial differential equations. They are first converted to dimensionless form, discretised and then solved using Runge-Kutta method of fourth order and Backward Implicit scheme (Chauhan and Srivastava, 2008a).

Results and Discussion

As HCs are most difficult to oxidize during the warm-up period, so their conversion is analysed in the converter channel. HCs propylene and propane are taken to represent total HCs and their oxidation in presence of platinum catalyst is analysed (Chauhan and Srivastava, 2008b). A mixture of propylene (1900 ppm) and propane (950 ppm) exhausts gases together at 390 °C enters a relatively cooler converter, initially maintained at 45 °C. Simultaneous changes in the gases concentration, gas temperature and solid temperature is observed for propylene and propane gases present in the exhaust. It was observed that initially when the converter is started, the solid temperature being very low, the start of catalytic reactions is not immediate. The reaction starts only after the incoming gases have heated the solid catalyst to its operating temperature level. Also, it is observed that the conversion of propylene in the two-gas system is more as compared to propane as shown in Figure 3.2 and it is only at higher catalyst temperatures that the conversion of propane becomes appreciable.

Figure 3.3 shows the comparison of gas and solid temperatures along the converter length with respect to time for a mixture of propylene and propane exhausts entering together at 390 °C over the solid present initially at ambient temperature. It is found that initially the gas heats the solid to bring it to its required operational temperature. Later due to onset of the exothermic reactions, solid temperature becomes higher than that of the gas phase and now heat is transferred from the solid to the gas phase.

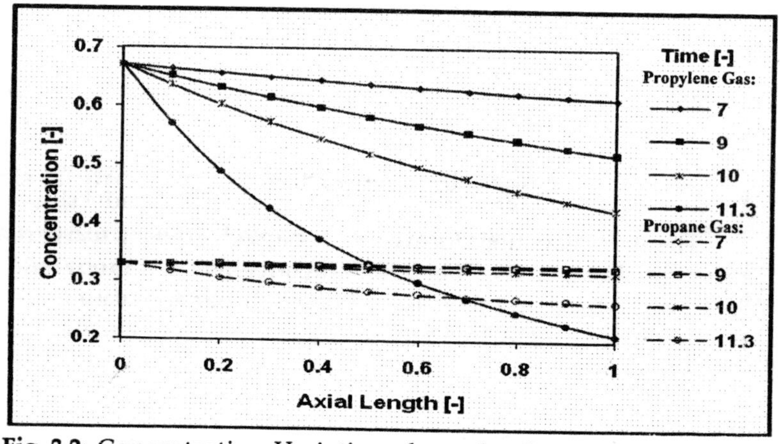

Fig. 3.2: Concentration Variation along the Converter Length for Propylene and Propane with Time.

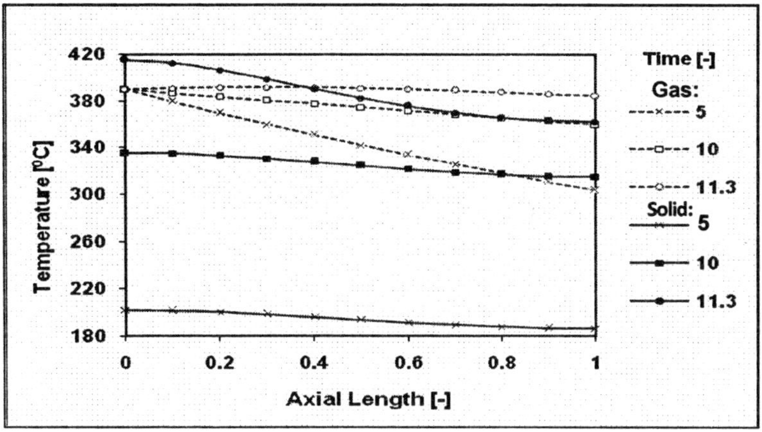

Fig. 3.3: Solid and Gas Temperature Variations along the Converter Length with Time.

In another work by Chauhan and Srivastava (2008c), the effect of changing the inlet gas temperature and solid temperature on warm-up period of the catalyst was analysed. In order to bring about a faster conversion of the pollutant propylene, the warm up period of the converter can be reduced by either introducing gas at higher temperature as shown in Figure 3.4 or by initially preheating the converter as shown in Figure 3.5 where the inlet gas temperature is kept constant at 390 °C and solid is taken at different initial temperatures of 20 °C, 45 °C, 100 °C, 200 °C, 300 °C, 327 °C and 350 °C.

In the case of preheated converter, the catalytic reactions start much earlier than in cases where the converter is at ambient temperatures. In the later case heating by the incoming hot gas has to be carried out for some time before the effective operating temperature is achieved. Computations based on the model show that for a preheated converter taken at 350 °C, the increase in temperature of incoming gas increases the conversion of propylene extremely fast as shown in Figure 3.6. So these results indicate the profound effect of solid catalyst heating in decreasing the warm up period considerably.

Conclusions

As experimental testing is expensive, development of a computer

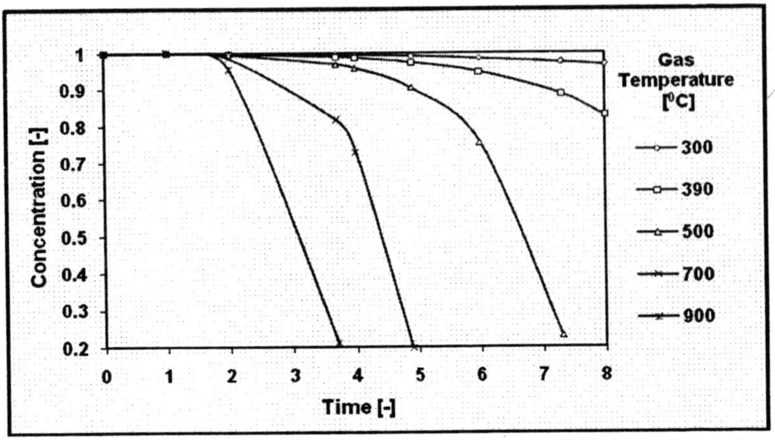

Fig. 3.4: Exit Concentration Variation of Gas for Different Inlet Gas Temperatures.

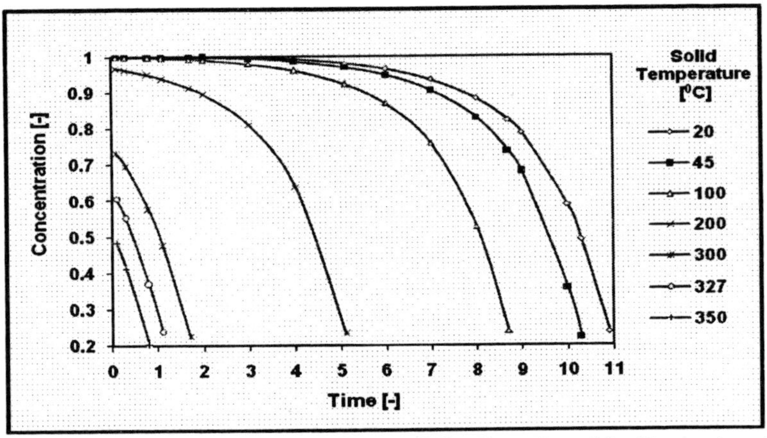

Fig. 3.5: Exit Concentration Variation of Gas Introduced in Converter at 390 °C with Time for Different Temperatures of Preheated Solid Catalyst.

model capable of simulating the components of an exhaust gas after-treatment system using one-dimensional model for predicting the hydrocarbon emissions during warm up period of the converter was carried out. As under normal conditions, it takes some time for the converter to heat up to its operating temperature, so during this warm-up period hydrocarbons are released untreated to the atmosphere. Analysis for decreasing this warm-

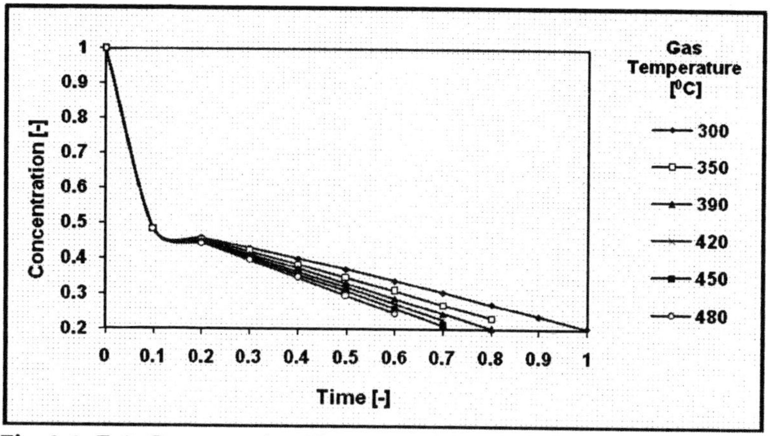

Fig. 3.6: Exit Concentration Variation of Gas in a Preheated Converter at 350 °C with Time for Different Inlet Gas Temperatures.

up period by increasing the inlet gas temperature or preheating the solid have been analysed individually and in combination. The later case is found to substantially decrease the warm-up period of the converter and hence decreases the release of pollutants to the atmosphere.

Nomenclature

a	Catalytic surface area per unit reactor volume, cm^2/cm^3
C	Concentration of the species, $gmole/cm^3$
Cp	Specific heat, cal/g K
d	Hydraulic diameter, cm
Ecat	Activation energy for the catalytic reaction, cal/gmole
K_g	Mass transfer coefficient, cm/s
Kcat	Pre-exponential factor for catalytic reaction, cm/s
L	Length of monolith, cm
R	Gas constant, cal/gmole K
S	Geometric surface area per unit reactor volume, cm^2/cm^3
T	Temperature, K
T	Time, s
V	Gas velocity, cm/s
X	Axial coordinates, cm
Greek letters:	
Λ	Thermal conductivity, cal/cm s K
ρ	Density, g/cm^3
Superscript:	
o	Initial conditions
Subscript:	
s	Solid
g	Gas
1	C_3H_6 gas
2	C_3H_8 gas

REFERENCES

Ahn, T., Pinczewski, W.V., and Trimm, D.L. (1986). "Transient Performance of Catalytic Combustors for Gas Turbine Applications." *Chem. Eng. Sci.*, 41, 55-64.

Aimard, F., Li, S. and Sorine, M. (1996). "Mathematical Modeling of Automotive Three-Way Catalytic Converters with Oxygen Storage Capacity." *Control Engg. Practice*, 4, 1119-1124.

Chauhan, S., and Srivastava, V.K. (2008a). "Modeling Exhaust Gas Pollution Abatement: Part I—Single Hydrocarbon Propylene." *Computers and Mathematics with Applications*, 55, 319-330.

Chauhan, S., and Srivastava, V.K. (2008b). "Modeling for Propylene and Propane present in the Exhaust Gases in a Monolithic Catalytic Converter." *Indian Chemical Engineer*, 50, 301-311.

Chauhan, S., and Srivastava, V.K. (2008c). "Modeling the Effect of Temperature on Propylene Conversion during Warm-up in a Monolithic Converter." *Chemical Product and Process Modeling*, 3(1) Article 43, 1-12.

Chen J., Yang H., Wang N., Ring Z. and Dabros T. (2008). "Mathematical Modeling of Monolith Catalysts and Reactors for Gas Phase Reactions." *Applied Catalysis A: General*, 345, 1-11.

Cooper, B.J. (2002). "The Catalytic Control of Motor Vehicle Emissions." *Commemorative Lecture at the Twenty-Third Honda Prize Awarding Ceremony*, Tokyo.

Dupont, V., Zhang, S.H., Bentley R. and Williams, A. (2002). "Experimental and Modeling studies of the Catalytic Combustion of Methane." *Fuel*, 81, 799-810.

Farrauto, R.J. and Heck, R.M. (1999). "Catalytic Converters: State of the Art and Perspectives." *Catalysis Today*, 51, 351-360.

Gandhi, H.S., Williamson, W.B. and Bomback, J.L. (1982). "Deactivation of Three-Way and Oxidation Catalyst Dual Bed Emission Control Systems: Catalyst Post Mortem Analyses from Methanol-Fueled Vehicles." *Applied Catalysis*, 3, 79-88.

Hayes, R.E., Kolaczkowski, S.T. and Thomas, W.J. (1992). "Finite-Element Model for a Catalytic Monolith Reactor." *Computers and Chemical Eng.*, 16, 645-657.

Kolaczkowski, S.T. (1999). "Modeling Catalytic Combustion in Monolith Reactors—Challenges Faced." *Catalysis Today*, 47, 209-218.

Koltsakis, G.C., Konstantinidis, P.A. and Stamatelos, A.M. (1997). "Development and Application Range of Mathematical Models for 3-way Catalytic Converters." *Applied Catalysis B: Environmental*, 12, 161-191.

Koltsakis, G.C. and Stamatelos, A.M. (1997). "Catalytic Automotive Exhaust Aftertreatment." *Prog. Energy Combust. Sci.*, 23, 1-39.

Montreuil, C.N., Williams, S.C. and Adamczyk, A.A. (1992). "Modeling Current Generation Catalytic Converters: Laboratory Experiments and Kinetic Parameter Optimization—Steady State Kinetics." *SAE Paper No. 920096,* 43-66.

Rijkeboer, R.C. (1991). "Catalysts on Cars—Practical Experience." *Catalysis Today,* 11, 141-150.

Shelef, M. and McCabe, R.W. (2000). "Twenty-Five Years after Introduction of Automotive Catalysts: What Next?" *Catalysis Today,* 62, 35-50.

Wanker, R., Raupenstrauch, H. and Staudinger, G. (2000). "A Fully Distributed Model for the Simulation of a Catalytic Combustor." *Chem. Eng. Sci.,* 55, 4709-4718.

4

Resource Management for Climate Resilience in Urban Areas

Meenakshi Dhote and *Kusum Lata*

Introduction

Urbanization as a phenomenon is gaining momentum at a rapid pace all over the world. In many parts of the world, especially in developed countries, a majority of the population lives in urban areas. Number of cities have also increased, as there were only 83 cities in 1950 with a population of not more than one million and 34 of them in the developing country, and today there are around 280 of such settlements in the world and are expected to double by the year 2015 (UN estimates). An attempt to view the human settlements as a resource (Natural, cultural, human, economic etc.) identifies the critical role played by the natural resource base to sustain the human population. So far, in cities, open spaces are supposed to mainly cater to the aesthetic demands and the recreation needs. But if we examine their environmental role, these are also many—pollution abatement, water recharge, indicator of pollution, climatic amelioration, flood control, etc. It is often felt that these functions are performed by plants. What we do not understand is that, the animals and birds form extremely important indicators of health of the environment of an area. In urban areas presence of certain birds, animals, and insects gives us clues of the type of habitat and its condition. Similarly, in open spaces in an urban area, if the species richness is high, it indicates the status of the habitat—age, condition of soil, chances of presence of other organisms and ultimately its fitness to withstand any change; which indicates sustainability of the environment.

Ecological Infrastructure of Urban Areas

An attempt to view human settlements as habitats or ecosystems identifies the critical role played by the natural resource base to sustain the human population. The various physical, biological, and chemical components (i.e. land, air, water, vegetation, animals, etc.) collectively contribute in regulating water availability, controlling climate, improving air quality, enhancing aesthetic potential etc. These resources can be termed as the ecological infrastructure of a settlement. These functions performed by them can be summarized as follows:

- **Regulation Functions:** Capacity of the natural/semi-natural ecosystem to perform the essential ecological process and life supporting system (providing clean air, water and soil, etc.)
- **Carrier Functions:** Capacity of the natural and semi-natural ecosystem by supporting many of the human activity by providing adequate space and medium.
- **Production Functions:** This is the capacity of the nature to produce the food and raw materials for human consumption and industrial use and to function as energy resource and genetic material.
- **Information Functions:** This is the capacity of nature to contribute to the spiritual enrichment, cognitive development, and aesthetic experience.

In the case of urban areas most of these functions are carried out by open spaces, green areas, water bodies and unpaved areas.

Urban open/ Green Spaces	Regulation Functions	• Regulates the chemical composition of atmosphere and purifies the local air.
		• Controls the run-off and flooding.
		• Regulates the hydrologic cycles.
		• Supports the biological diversity in the city.
		• Prevents soil erosion and sedimentation.
		• Regulates the local and global climate.

(*Contd.*)

	Carrier Functions	• Provides food and raw materials.
		• Conserves the energy in the city through controlling the micro climatic variations.
		• Provides recreation and tourism.
		• Integrates urban man to the nature.
	Production Functions	• Produces oxygen.
		• Recharges the groundwater tables.
		• Provides medical resources.
		• Produces raw materials for some of the human activity.
	Information Functions	• Aesthetic information.
		• Spiritual and religious information.
		• Cultural and artistic inspiration.
		• Scientific and educational information source.

Urban Habitats

The habitat diversity or biodiversity of an urban area is reflected in a greater number of species than in a comparably sized area of more homogeneous non-urban use. Urban biodiversity is enhanced by successful hybridization and deliberate introduction. Any consideration of the urban ecosystem should be able to identify the area, form and pattern that supports a particular diversity. These variations are a function of age, location, and the physical and cultural environment in which the city has developed.

The city is a natural system made artificial, the functioning of which requires the presence of certain number of different species. The human species is not the only one living in cities: there are always either remnants of nature, which preceded the city, or biotic exotic elements introduced by man. Controlling and conserving these elements, improving and increasing their numbers, is a primary objective of cities, for scientific as well as social reasons. Urban biodiversity can be categorised as natural, semi-natural and artificial habitats (*see* Figure 4.1).

The Urban Ecological Gradient

"Urbanization results in a complex pattern of intermixed high and low-density built up areas. Alternative urban development patterns affect ecological processes both directly—by replacing

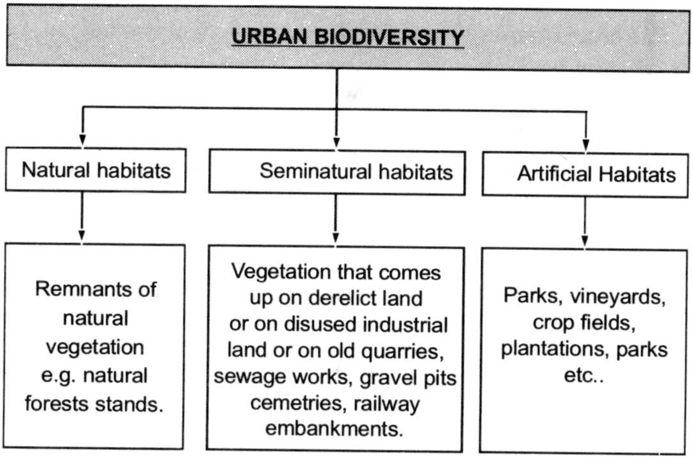

Fig. 4.1: Categories of Urban Habitats.

native habitat with simplified human-dominated systems and indirectly by rearranging the biophysical attributes that cause a variety of interrelated local and global effects. Ecologists often have treated "urban" as a separate type of land-cover, akin to "agriculture" or "grassland". In fact, "urban" is an endpoint on a continuum of settlement intensity—a gradient that extends from sparsely populated rural areas to large cities. When we compare densely populated urban areas to suburban development, there is a noticeable increase in average lot size (and thus a decrease in human density), the amount of green space, and the density of trees and shrubs. Suburban areas often give way to exurban residential development that is characterized by still larger lot sizes and greater amounts of vegetation. Ultimately, exurban development grades to rural area, dominated by agricultural activities, or wildlands. Settlement is not a matter of "either—or" but one of degree." An important advantage of the gradient approach, lies in its potential for identifying threshold or breakpoints where human impacts cause marked changes in biotic response. Such threshold might be found, for example, in measures of human density, patterns of settlement, or distance from developed areas. Thresholds are important simply because they offer a clearer basis for policy, such as the determination of buffer distances for sensitive species that occur in particular habitats or appropriate zoning restrictions.

Resource Management for Climate Resilience in the Urban Areas

The future city spatial patterns, growth and development are prove to be affected by the climate change. As in ecology, resilience has been described as the capacity of an ecosystem to tolerate disturbance without collapsing into a qualitatively different state. To become more resilient, cities will need to adopt planning and design strategies that allow them to develop the capacities to better respond and adapt to the economic, social and physical stresses they will face as they confront the twin challenges of resource depletion and climate change (*see* Figs. 4.2 and 4.3).

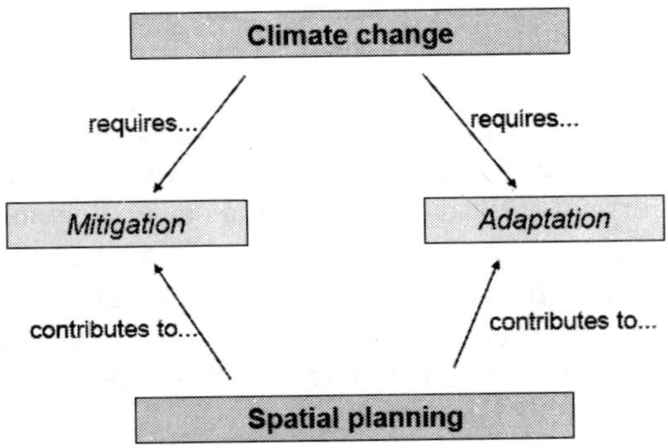

Fig. 4.2

Source: Spatial Planning as an Adaptation Strategy to Climate Change, *Mark Fleischhauer.*

Planning Response to Climate Change (*see* Fig. 4.4)

- Land-use planning should focus on efficient use of resources, energy consumption, and infrastructure.
- Compact city or sprawl city.
- Reducing Greenhouse Gas Emissions through Transportation Planning for reducing transportation related greenhouse gas emissions is to reduce vehicle trips through mixed-use.
- compact developments.

Projected change in extreme climate to phenomena and their likelihood	*Consequences of climate change*
Warmer and fewer cold days and nights, warmer and more frequent hot days and nights (virtually certain)	• Heat island effect
	• Increased demand for cooling
	• Declining air quality in cities
	• Effects on winter tourism
	• Reduced energy demand for heating (a short-term benefit but not in East Asia)
	• Reduced disruption to transport due to snow, ice (a short-term benefit but not in East Asia)
Warm spells/heat waves. Frequency increases over most land areas (very likely)	• Increased water demand
	• Water quality problems
	• Increased risk of heat-related mortality, especially for the elderly, chronically sick, very young and socially isolated
	• Reduction in quality of life for people in warm areas without appropriate housing
Heavy precipitation events. Frequency increases over most areas (very likely)	• Adverse effects on quality of surface and groundwater
	• Contamination of water supply
	• Increased risk of deaths, injuries, infectious, respiratory and skin diseases
	• Disruption of settlements, commerce, transport, and societies due to flooding
	• Large displacement of people
	• Pressures on urban and rural infrastructures
	• Loss of property
	• Water stress may be relieved (short-term benefit)
Intense tropical cyclone activity increases (likely)	• Power outages
	• Distress migration to urban areas
	• Disruption of public water supply
	• Increased risk of deaths, injuries, water and food-borne diseases; and post-traumatic stress disorders

(*Contd.*)

Fig. 3: (*Contd.*)

	• Disruption by flood and high winds
	• Withdrawal of risk coverage in vulnerable areas by private insurers
	• Potential for population migrations
	• Loss of property
Increased incidence of extreme high sea level (excludes tsunamis) (likely)	• Decreased freshwater availability due to salt-water intrusion
	• Increased risks of deaths and injuries by drowning in floods and migration-related health effects.
	• Loss of property and livelihood
	• Permanent erosion and submersion of land
	• Costs of coastal protection versus costs of land-use relocation
	• Potential for movement of population and infrastructure

Fig. 4.3: Climate Change Impact on Urban Areas.

Source: IPCC, *Synthesis Report—Summary for Policymakers*. Assessment of Working Groups I, II, and III to the Third Assessment Report of the International Panel on Climate Change (IPCC: Cambridge University Press, 2007.

- Reducing Greenhouse Gas Emissions through Building Siting and Design.
- Energy consumption in buildings is a result of building design and siting, building operations, the urban form.

NAPCC and Resource Management in Urban Areas

NAPCC (the National Action Plan on Climate Change) with its eight missions provides a framework for addressing climate change as a core development issue. The Prime Minister of India released the country's NAPCC (National Action Plan on Climate Change) on 30 June, 2008. This plan essentially pursues the objectives of sustainable development that would increasingly guide India's development policy in the future. The plan also responds effectively to the growing evidence of global climate change, particularly as brought out in the Fourth Assessment

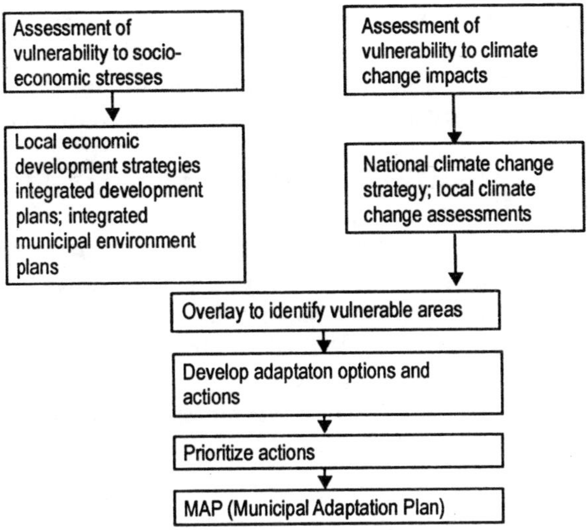

Fig. 4.4: Process for Developing a Municipal Adaptation Plan (MAP).

Source: Developing a Adaptation Plan Pierre Mukheiber and Gina Ziervogel

Report of the IPCC (Intergovernmental Panel on Climate Change).

The eight missions mentioned in the NAPCC map out long-term and integrated strategies to achieve key national goals from the climate change perspective. The NAPCC identifies measures that promote development objectives of the country while producing co-benefits that address climate change effectively.

National Missions

National Solar Mission: The NAPCC aims to promote the development and use of solar energy for power generation and other uses with the ultimate objective of making solar competitive with fossil-based energy options. The plan includes:

- Specific goals for increasing use of solar thermal technologies in urban areas, industry, and commercial establishments;
- A goal of increasing production of photovoltaics to 1000 MW/year; and
- A goal of deploying at least 1000 MW of solar thermal power generation. Other objectives include, the

establishment of a solar research centre, increased international collaboration on technology development, strengthening of domestic manufacturing capacity, and increased government funding and international support.

National Mission for Enhanced Energy Efficiency: Consequent to the announcement of the National Action Plan on Climate Change in June 2008, which identified development of solar entergy technologies in the country as a National Mission, the Government of India has approved "Jawaharlal Nehru National Solar Mission" (JNNSM)[1] which aims at development and deployment of solar energy technologies in the country to achieve parity with grid power tariff by 2022. The objective of the National Solar Mission is to establish India as a global leader in solar energy, by creating the policy conditions for its diffusion across the country as quickly as possible. The Mission will adopt a 3-phase approach, spanning the remaining period of the 11th Plan and first year of the 12th Plan (up to 2012-13) as Phase 1, the remaining 4 years of the 12th Plan (2013-17) as Phase 2 and the 13th Plan (2017-22) as Phase 3. At the end of each plan, and mid-term during the 12th and 13th Plans, there will be an evalution of progress, review of capacity and targets for sbsequent phases, based on emerging cost and technology trends, both domestic and global. The aim would be to protect Government from subsidy exposure in case expected cost reduction does not materialize or is more rapid than expected. The immediate aim of the Mission is to focus on setting up an enabling environment for solar technology penetration in the country both at a centralized and decentralized level. The first phase (up to 2012-2013) focused on capturing of the low-hanging options in solar thermal; on promoting off-grid systems to serve populations without access to commercial energy and modest capacity addition in grid-based systems. In the second phase, after taking into account the experience of the initial years, capacity was aggressively ramped up to create conditions for up scaled and competitive solar energy penetration in the country.

National Mission on Sustainable Habitat: To promote energy efficiency as a core component of urban planning, the plan calls for:

- Extending the existing Energy Conservation Building Code;

- A greater emphasis on urban waste management and recycling, including power production from waste;
- Strengthening the enforcement of automotive fuel economy standards and using pricing measures to encourage the purchase of efficient vehicles; and
- Incentives for the use of public transportation.

National Water Mission: With water scarcity projected to worsen as a result of climate change, the plan sets a goal of a 20 per cent improvement in water use efficiency through pricing and other measures.

National Mission for Sustaining the Himalayan Ecosystem: The plan aims to conserve biodiversity, forest cover, and other ecological values in the Himalayan region, where glaciers that are a major source of India's water supply are projected to recede as a result of global warming.

National Mission for a "Green India": Goals include the afforestation of 6 million hectares of degraded forest lands and expanding forest cover from 23 per cent to 33 per cent of India's territory.

National Mission for Sustainable Agriculture: The plan aims to support climate adaptation in agriculture through the development of climate-resilient crops, expansion of weather insurance mechanisms, and agricultural practices.

National Mission on Strategic Knowledge for Climate Change: To gain a better understanding of climate science, impacts and challenges, the plan envisions a new Climate Science Research Fund, improved climate modeling, and increased international collaboration. It also encourages private sector initiatives to develop adaptation and mitigation technologies through venture capital funds.

Some of the Missions are directly addressing resources like National Solar and Water Missions. Other Missions are involved indirectly into resource management, i.e. Mission on Sustaining the Himalayan Ecosystem which focus of biodiversity conservation, The Green India Mission which deals with the forest resource.

Implementation

Ministries with lead responsibility for each of the missions are

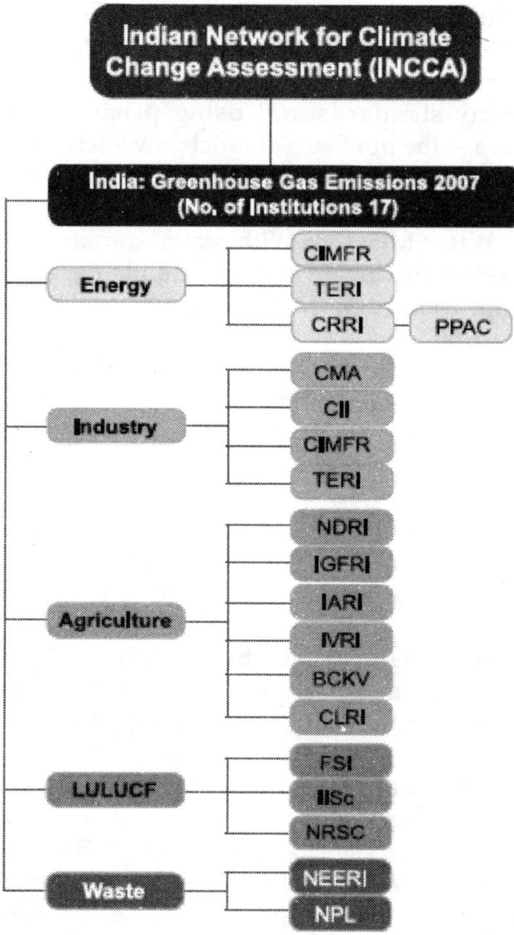

Fig. 4. 5

directed to develop objectives, implementation strategies, timelines, and monitoring and evaluation criteria, to be submitted to the Prime Minister's Council on Climate Change. The Council will also be responsible for periodically reviewing and reporting on each mission's progress. To be able to quantify progress, appropriate indicators and methodologies will be developed to assess both avoided emissions and adaptation benefits (*see* Fig. 4.5).

Organizations Associated with Indian Network for Climate Change Assessment (INCCA)

Indian Network for Climate Change Assessment, the Network comprises of 127 institutions and 228 (INCCA) scientists across India and was launched on October 14, 2009. Its role comprises of—

- Assessing the drivers and implications of climate change through scientific research.
- Preparing climate change assessments once every two years (GHG estimations and impacts of climate change, associated vulnerabilities and adaptation).
- Developing decision support systems.
- Building capacity towards management of climate change related risks and opportunities.

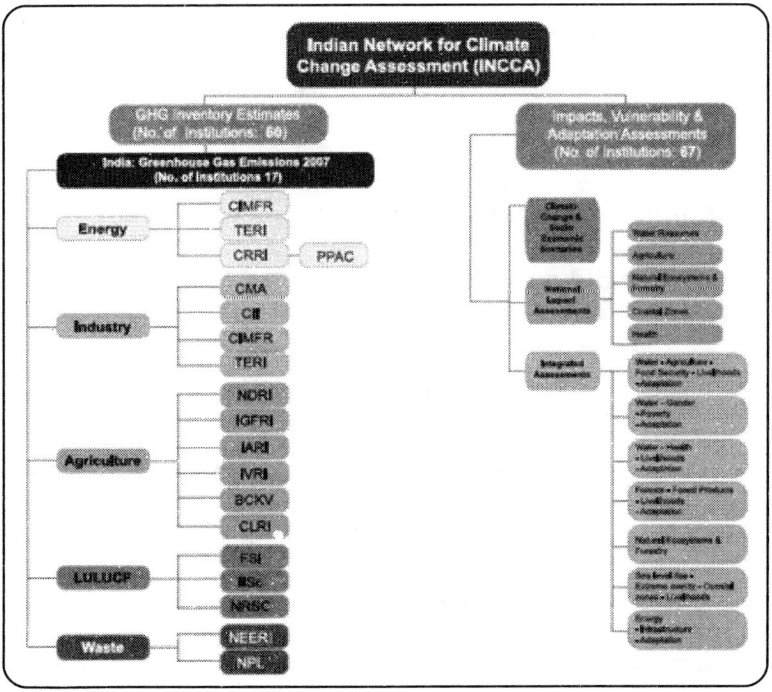

Fig. 4.6: Institutions involved preparing India's Green House Gas Emission.

The 17 institutions which were primarily involved in preparing India's Green House Gas Emission are:

BCKV	Bidhan Chandra Krishi Viswavidyalaya
CII	Confederation of Indian Industry
CIMFR	Central Institute of Mining and Fuel Research
CLRI	Central Leather Research Institute
CMA	Cement Manufacturing Association
CRRI	Central Road Research Institute
FSI	Forest Survey of India
IARI	Indian Agricultural Research Institute
IGFRI	Indian Grass and Fodder Research Institute
IISc	Indian Institute of Science
IVRI	Indian Veterinary Research Institute
NDRI	National Dairy Research Institute
NEERI	National Environment Engineering Research Institute
NPL	National Physical Laboratory
NRCS	National Remote Sensing Centre
PPAC	Petroleum Planning and Analysis Cell
TERI	The Energy and Resources Institute.

Institutional Capacity Building

Capacity building is essential at two levels—one is building capacity at the institutional level and the other is at the individual level. Building capacity at the institutional level is essential to address the needs of inventory preparation for various purposes, be it at national scale, sectoral scale or at installation level. Therefore, there is a need to have a National Inventory Management System. This will also include additional involvement of institutions with varied research experience to widen the pool of institutions that will look at the various aspects of inventory development.

The individual capacity building is another aspect that needs improvement of the skills of individuals on a continuous basis to be in consonance with the latest developments in the process and subject of inventory preparation for various sectors, including Energy, Industry, Agriculture, Land Use, land use change and forests, and waste management.

Establishment of National Inventory Management System

Building on the base of knowledge institutions already engaged in the preparation of a national assessment a National Inventory Management System (NIMS) under the MoEF can be developed. The NIMS may address the requirements, of documentation, archiving and continuous updating of the databases as well as the QA/QC and uncertainty management issues of the GHG inventories being developed across the years. The NIMS mandate should address the development of Systemic tools and procedures for documenting methodologies, creation of databases of emission factors and activity data for each point source and the various disaggregated sources that add up to generate the national GHG Emission profiles across each year. The NIMS mandate may include:

- Undertake data management and collection on an annual basis;
- Devise strategies for data generation and improvement;
- Establish systems for data archiving and record keeping;
- Ensure mechanisms for synchronization and cross-feeding between emission inventories, national energy balances and relevant sector surveys; and
- Provide guidance for technical peer reviews, procedures for QA/QC and uncertainty management.

A web-based data base management system may further help in wider accessibility of data to concerned stakeholders and also towards visualisation of GHG data thus generated.

Enhanced Networking and Individual Capacity Building

The present network of institutions involved in this assessment is only estimating the Green House Gas (GHG) emissions by sector. Green House Gas (GHG) inventory preparation is a much larger exercise, as it involves data generation, collection, archival, literature survey for emission factors, determination of country specific emission factors, undertaking QA/QC and uncertainty analysis amongst other activities. For each of these skilled approach is required and therefore trained manpower is necessary that will also improve methodologies of estimation with respect to the International methodologies available and may develop new methodologies to reflect the national circumstances of specific activities that are key categories in the entire Green House Gas (GHG) inventory. This can be done by enlarging the base of institutions involved in this present activity which can skillfully steer each and every activity required for Green House Gas (GHG) inventory preparation. Other than the enhancement in the institutional base, it is also important to train manpower in the latest techniques of inventory preparation as well as bring in new capacity to carry forward the work.

Efforts at Urban Area Level—Delhi

Delhi has a complex administrative set-up as it is the seat of the Central and State Governments. The administrative setup of Delhi has undergone changes from being a State till 1956 to Union Territory and finally to Government of National Capital Territory Delhi in 1992. The Legislative Assembly of Delhi has powers to make laws in respect of all the 68 matters in the State list excepting public sector, police and land. The Municipal Corporation of Delhi (MCD) has jurisdiction over almost the entire capital territory including rural areas, but excluding New Delhi area which has New Delhi Municipal Council (NDMC) and Delhi Cantonment area which has Delhi Cantonment Board. There are perennial interventions of several public sector bodies in the management of the Capital City. The Delhi Development Authority is the major actor in the field of urban planning; land-use planning and management; and housing. Further there are about 120 public bodies intervening on the same territory and serve the community in one aspect or the other.

The organization providing the services related to Land-use Planning and Urban Development is Delhi Development Authority (DDA) while the planning at regional level is being done by the National Capital Region Planning Board (NCRPB). The organizations providing the services related to land are Central Government, DDA, MCD, NDMC, Cantt. Board, PWD, Delhi Government, DSIDC, Cooperatives/Private Organisations providing Enforcement of Planning Controls are Special Task Force (Home Ministry), DDA, MCD, NDMC, Cantt. Board. There is a lack of clarity over departmental responsibilities for land use planning, development, maintenance and enforcement. The geographical boundary of the State Government and MCD are

Table 4.1: Major Organisations Providing Services in the NCT of Delhi

S. No.	*Service*	*Organisation providing the service*
1.	Urban Planning and Development	Delhi Development Authority (Also NCRPB) (MoUD)
2.	Land	Central Govt., DDA, MCD, PWD, Delhi Govt., DSIDC, Cooperatives/Private, etc.
3.	Enforcement of Planning Controls	Special Task Force (Home Ministry) D.D.A., MCD, NDMC
4.	Urban Form and Aesthetics	Delhi Urban Art Commission
5.	Slum/Jhuggi Jhopri rehabilitation	Delhi Slum Improvement Board (Delhi Government) Slum & JJ Deptt. (MCD)
6.	Recreation	DDA, MCD, NDMC, TFAI, Private Sector
7.	Lands, Residential and industrial	DDA, CPWD, DSIDC Govt. of Delhi, MCD
8.	Horticulture	DDA, MCD, NDMC, CPWD
9.	Heritage Conservation	Archaeological Survey of India, GNCTD, DDA, INTACH
10.	Building Plan Approval	MCD, DDA, NDMC, DUAC, AAI, DPCC, DFS, etc.
11.	Pollution Control	Conservator of Forests, Delhi Govt., Env. Pollution Control Authority, DOE, Delhi Pollution Control Committee, CPCB, MCD
12.	Drainage and Flood Control	I&F Department, GNCTD, DDA, MCD
13.	Water and Sewerage	Delhi Jal Board (Delhi Govt.), MCD, NDMC (Autonomous elected body under GoI,)
14.	Electricity	Delhi Vidyut Board of Delhi Government on July 01, 2002 was unbundled into six successor companies: Delhi Power Supply Company Limited (DPCL)-Holding Company; Delhi Transco Limited (DTL)—TRANSCO; Indraprastha Power Generation Company Limited (IPGCL)—GENCO; BSES Rajdhani Power Limited (BRPL)—DISCOM; BSES Yamuna Power Limited (BYPL)—DISCOM; North Delhi Power Limited (NDPL)—DISCOM.
15.	Transport	Delhi Transport Corporation, State Tpt. Corpn., Pvt. Operators
16.	Roads	Central Govt. (N.H.), PWD (Ring Road), MCD, NDMC, Cantt. Board., DDA (Inner city Roads)
17.	Education	Central Govt./State Govt.

(*Contd.*)

S. No.	Service	Organisation providing the service
18.	Universities, Colleges	MCD, NDMC, Govt. of National Capital Territory of Delhi (GNCTD), Private
19.	Health	MCD, NDMC, GNCTD, Private
20.	Utilities	DDA, MCD, NDMC, Delhi Govt., etc.
21.	Commercial	DDA, NDMC, MCD, DAMB, L & DO, Govt. of India
22.	Security and Emergency Services	Delhi Police, Delhi Fire Service
23.	Fuel Distribution	Oil Companies, Gas Authority of India
24.	Safety	Fire Department, Boiler Inspector, MCD, DOI, Central Government (Department. of Explosives) etc.
25.	Telephones, Cellular, Internet	MTNL, Private companies, VSNL, etc.
26.	Sales Tax Registration	Govt. of Delhi
27.	Factory License	MCD, DOI, CIF, Labour Department., Delhi Govt.
28.	Milk	Delhi Milk Scheme, Mother Dairy (Delhi Govt./Central Govt.)

coterminus, and their functions are almost the same. Of the below mentioned 28 services, top 12 services are directly/indirectly related to Climate resilience in the NCT of Delhi. Within these 12 services, the mandate of the organisations providing these services varies from policy formulation, planning, implementation to management. Their area of jurisdiction also varies and there is not any coordination amongst them (*see* Table 4.1).

Delhi's Chief Secretary Shri Rakesh Mehta sets the Agenda for Delhi through a monograph titled "Climate Change: Agenda for Delhi 2009-2012". This monograph starts with Mahatma Gandhi's dictum that the "earth has enough resources to meet people's needs, but will never have enough to satisfy people's greed". Hence, it has been rightly pointed out that "defining needs and meeting them while avoiding greed at all costs must be our defining agenda". Climate Change is a global challenge and Delhi also has to contribute to the mitigation of the adverse effects of Climate Change. Hence, the agenda for Climate Change has to be integrated into the policy framework of Delhi's development in the future. A positive development policy response to Climate Change is required to ensure that the objectives of growth with equity are achieved while at the same time, ensure long term sustainability of the environment. This can be achieved by preparing an operational plan which gets reflected in every department which has related works to be undertaken.

The 2030 Agenda for Sustainable Development Goal, The new universal Agenda as approved by United Nations is a "Plan of Action for People, Planet and Prosperity" adopted by many member countries for implementation. India as a member country of the United Nations and signatory to UN Sustainable Development Summit in September 2015 has adopted the SDG 2030 as the guiding framework for its National Development Agenda till 2030. The prime objective of this universal agenda is to eradicate poverty in all its forms and dimensions including extreme poverty that is the greatest global challenge and an indispensable requirement for sustainable development. The SDG 2030 includes 17 Goals with 169 targets. The SDG framework is built on the millennium development goals and aims to realize the human rights of all and to achieve gender equality and empowerment of women by stimulating action in the areas of critical importance of humanity which is economic, social and environmental. Government of Delhi also aspires to provide environmentally resilient habitat with adaptive capacity to climate change, sustainable management to terestrial eco-systems and forests.

For mega city Delhi, ensuring Sustainable Habitat is one of the most important challenges. In order to make the implementation of the Climate Change Mission on Sustainable Habitat a reality, the promulgation of modern building byelaws is a must. The adoption of the Green Building Code is a good starting point and the star rating of electrical equipments is a good message to use energy efficient appliances and technology for the purpose. There is also a need to promote standardization of construction materials incorporating various elements of green building code in it. The need is to create a focus group to address the manifold issues in this area.

Green cover area in Delhi is 20 per cent of total area, which is much larger than in other cities, hence Delhi is also known as one of the greenest cities of the world. Delhi has taken the lead in greening with the effective green cover increasing from 26 sq kms to 300 sq kms in ten years. This is proposed to be increased to 33 per cent over time as per the target set out in the National Green Mission. This is proposed to be done by a series of measures which includes more effective greening of existing greens, opening of new city forests and intensifying the green cover over the entire ridge area of Delhi which covers nearly 6000 hectares. Each department has a target to fulfil and this is checked by the forest department. The setting up of the Delhi Parks and Gardens Society is meant to coordinate the greening efforts to reach the target figure of 33 per cent.

In order to tackle the problem of climate change effectively in an urban environment, there is a need to evolve effective policies which can be reflected in the development plans along with relevant technologies. This needs an effective knowledge base and real time management of the processes. Delhi has taken the lead in identifying the Mahatma Gandhi Institute of Climate Change Challenge in Bakoli as the field level institution which will pioneer efforts in creating strategic knowledge about climate change. To make this possible, the Energy Efficiency and Renewable Energy Management Centre has been placed under the MGICCC to coordinate the activities of climate change and energy management in a holistic manner. The Centre will partner with leading institutions in creating awareness and conducting research in this area.

NOTES

1. Government of India, Ministry of New and Renewable Energy, Jawaharlal Nehru National Solar Mission, Website: mnre.gov.in/ resoulution.
2. Jawaharlal Nehru National Solar Mission, towards, building Solar India, website: mnre.gov.in/sites/default/files/uploads/mission document JNNSM.
3. *Reconceptualising Smart Cities: A Reference Framwork for India,* Shrimoyee Bhattacharya, Sujay Rathi *et al.* (2015).

REFERENCES

American Planning Association Policy Guide on Energy (2003). <http://www.planning.org/policyguides/energy.htm> [6 July 2007]; Robert Riddell, *Sustainable Urban Planning: Tippling the Balance*. Malden, MA: Blackwell Publishing, 2004, 102; P.D. Fleming and P.H. Webber, "Local and Regional Greenhouse Gas Management", *Energy Policy*, 32 (2004) 763; Paul Parkern, Ian H. Rowlands, and Daniel Scott, "Innovations to Reduce Residential Energy Use and Carbon Emissions: An Integrated Approach", *Canadian Geographer*, 47, No. 2 (2003): 170.

Donley, Roarke, Bowman, Reed, Marzluff, John, M. (Eds.) (2001). *Avian Ecology and Conservation in an Urbanizing World*, Kluwer Academic Publishers, Dordrecht, Netherlands.

Ibrahim Dincer and Marc A., Rosen. "A Worldwide Perspective on Energy, Environment and Sustainable Development", *International Journal of Energy Research*, 22 (1998): 1305-1320.

IIRS, TC; Biodiversity and Environment; Geographic Information System Perspectives; IIRS, 2000,

Koen Steemers, "Energy and the City: Density, Buildings and Transport", Energy and Buildings, 35 (2003):13; Naess, 508; Erling Holden and Ingrid T. Norland, "Three Challenges for the Compact City as a Sustainable Urban Form", *Urban Studies*, 42, No. 12 (2005): 214.

Mehta, Rakesh Jan. 2009, Monograph titled "Climate Change, Agenda for Delhi 2009-2012", Directorate of Information and Publicity, Delhi Government.

Reinventing Planning: New Governance Paradigm for Managing Human Settlements, a Position Paper developing themes from the Draft Vancouver Declaration for debate leading into the World Planners

Congress in Vancouver, June 17-20, 2006, p. 2. <http://www.planning.org/knowledge/reinventingplanning.htm> [26 June 2006]; Worldwatch Institute, *2007* State of the World: Our Urban Future (New York, NY: W.W. Norton, 2007), 91; Edward J. Jepson, Jr., "The Adoption of Sustainable Development Policies and Techniques in U.S. Cities: How Wide, How Deep, and What Role for Planners?" *Journal of Planning Education and Research,* 23 (2004): 230.

Sundaram, A. Meenatchi; City Green Plan as a means to achieve sustainable development; (Background paper); Map India 2002, New Delhi, February 2002.

Tivy Joy (1993). *Bio Geography: A study of Plants in the Ecosphere,* Third Edition; Longman Scientific & Technical, 1993.

5

Impacts of Dams on Climatic Conditions and Agriculture of Hills

A Case Study of Chamera-I Hydroelectric Power Project in Ravi Basin of Himachal Pradesh

Mohinder Slariya

Introduction

Water determines every aspect of human life. Almost every civilization of the world has been developed beside the rivers. Historically speaking, all the settlements whether small or big have been developed beside the river or where the water remained available round the year. The examples of existence of any city throughout the world can be cited to support this argument. To a greater extent, the first civilizations in history appeared in the vicinity of water, such as Mesopotamia in Anatolia and the river Nile in Africa, Indus Valley civilization in India and led their lives in a very smooth and respectful manner by establishing the grounds of agriculture, trade, commerce and science. With the increase in population from earlier civilization till date water has become the single most precious and essential item that sustains life in this world. To meet out the ever increasing need of energy to run the industrial set up of any economy, this resource has drawn the attention of the policy-makers, planners and developers. But unfortunately because of rapid industrialization and unconscious and unscientific utilization of this natural resource, a threat on the lives and livelihood of people have been posed.

This resource has got special attention at international levels. The year 2009 has marked the halfway point of the 2005-2015 UN

Decade of Water for Life and has been recorded in history as a year during which all the progress made on water-related topics since the Rio Conference, Agenda 21, UN Millennium Development Goals and the 2002 Johannesburg Plan of Implementation have been reviewed and new initiatives been discussed.

After independence, with an objective to achieve rapid economic growth, a number of major developmental projects have been planned and implemented by both centre as well as state governments. These projects include construction of mega irrigation dams, power projects, industries, mining operation etc. Undoubtedly, these projects have provided irrigation facilities to thirsty lands, energy for growing industries and contributed a lot in the development of growing economy. Along with development, these mega projects have led to forced displacement of millions of people and numerous environmental and social problems, which pose threat to the existence of human beings and natural resources. Various case studies conducted by scholars reveal that the state has not taken the issue of displacement seriously and as many as 75 per cent of the displaced people have not been rehabilitated and their income and livelihood have not been restored properly (Fernandes 1991, Fernandes *et al.*, 1989).

Concept of Development

Initially, development meant fulfilment of the basic needs to the optimum level, but in the present context development means construction of multistory buildings, industries and construction of dams. According to United Nations Report (1982) *"Development refers to an overall process of transforming men and society leading to social order in which every human being can achieve moral and material well-being"*. Development affects all sections of the society. Each section and each institution i.e. social, economic, religious and political is affected by development. Although the effects may be less in some section and more in others. According to Collins Dictionary of Sociology, "Development is any type of change which results in increased economic productivity and prosperity and new and more complex forms of social structure and organization." In contemporary works, economic and social development is usually used to refer to "the process of industrialization in both its socialist and capitalist forms."

Hydroelectric Power Projects in India

Over the last two centuries, human activity has transformed the chemistry of Earth's water and air, altered the face of Earth itself, and rewoven the web of life. As far as the installation of hydroelectric power projects is concerned, the first hydropower project in India, of 4 x 200 kilowatts (kW) installed capacity, located at Darjeeling was installed way back in 1897 (GoI, 1972: v. 3, pt. 2:324). There has been impressive growth in the power sector in the last five decades of planned development. India's hydropower potential has been estimated by the Central Electricity Authority (CEA) as over 84,000 MW at 60 per cent load factor or 1,48,700 MW installed capacity. The details of major basin-wise hydroelectric power potential assessed by the CEA and the probable hydropower installed capacity is shown in Table 5.1.

Table 5.1: Hydroelectric Power Potential in India

S. No.	*River Basin(s)*	*Potential at 60 per cent Load Factor (MW)*	*Potential generation capacity (in MW)*
1.	Indus Basin	19,988	33,842
2.	Ganga Basin	10,715	20,711
3.	Brahmaputra Basin	34,920	66,005
4.	Central Indian Rivers System	2,740	4,152
5.	West flowing Rivers of South India	6,149	9,418
6.	East flowing Rivers of South India	9,532	14,511
	Total	**84,044**	**1,48,699***

*small/mini/micro power projects excluded.
Source: CEA, Government of India.

The small-scale hydel schemes in India have been categorised as micro (up to 180 kW), mini (up to 2 MW) and small (up to 15 MW). The total potential of small-scale hydel projects in India has been broadly assessed as 10,000 MW. The CEA had in 1998-99 assessed the small hydropower potential of about 6,000 MW capacity based on 1604 identified sites.

Pressure of Hydro-Power Generation in Himachal Pradesh

The hydro-electric power potential in Himachal Pradesh is

estimated at 25,000 MW, which is 24.90 per cent of India's total potential. Out of this, 6,045 MW [24.18%] has been harnessed so far, 2720.5 MW [10.84%] is under execution. Techno-economic feasibility studies are complete for 3,011 MW and in the process of completion for 3,671.5 MW. Survey has been completed for 4187 MW (Table 5.2).

Hydroelectric power development in Himachal Pradesh has got top priority from Sixth Five Year Plan i.e. 1980-85 onwards. During the Tenth Five Year Plan i.e. 2002-2007 besides completing the ongoing projects as early as possible, the state government has prepared an ambitious plan to accelerate hydro-power generation of identified potential capacity.

Table 5.2: Basin-wise Distribution of Hydel Capacity (in MW)

Basin	*Currently operational*	*Under execution*	*TEFR ready*	*TEFR not ready*	*Survey completed*	*Total*
Yamuna	0211.52	0110.00	0231.00	0000.00	0039.00	**0591.52**
Satluj	3150.25	1280.50	1402.00	2227.00	1360.50	**9420.25**
Beas	1634.50	1330.00	0736.00	0856.50	0025.00	**4582.00**
Ravi	1043.50	0000.00	0642.50	0348.00	0260.00	**2294.00**
Chenab	0005.30	0000.00	0000.00	0240.00	2503.00	**2748.30**
Total	**6045.07**	**2720.50**	**3011.50**	**3671.50**	**4187.50**	**19636.07***

*753 MW small/micro/mini hydel power potential have been excluded.
Source: Author's compilation from State Electricity Board Data (2007).

To achieve the targeted development, three major basins i.e. Satluj, Beas and Ravi, whose potential generation capacity is more than 72 per cent of total identified potential capacity, has been heavily targeted for the hydro-power generation. Ravi, Satluj and Beas have more than 72 per cent potential generation capacity and these three basins have been heavily targeted for the hydroelectric power generation (Table 5.2).

As shown in the Table 5.2 and 5.3, the Satluj basin is targeted for heaviest exploitation with 9,420 MW projects spread over 37 locations. Beas basin comes next with 4,582 MW, spread over 26 locations. Ravi and Chenab basins account for 5,042 MW spread over 46 locations together. Yamuna basin straddles two states [Himachal and Uttarakhand] and accounts for 591.5 MW spread over 12 locations. Thus, 45 projects are either operational or under

Table 5.3: Basin-wise Distribution/Status of Power Projects

Basin	Currently Operational	Under Execution	TEFR Ready	TEFR Not Ready	Survey Completed	Total
Yamuna	04	01	06	00	01	12
Satluj	09	05	02	08	13	37
Beas	08	06	08	03	01	26
Ravi	07	00	07	04	02	20
Chenab	05	00	00	01	14	20
Total	**33**	**12**	**23**	**16**	**31**	**115***

* In addition to these power projects, 286 Power Projects of different Magnitude are there in Himachal Pradesh.

Source: Author's compilation from State Govt. Data (2007).

execution. Out of 115 projects, 23.5 per cent [27] are large [over 200 MW], 37.4 per cent [43] are medium sized [50-200 MW] and 45 are of under 50 MW capacity. Satluj, Beas and Ravi basins account for 72.17 per cent medium and large projects.

Hydroelectric Power Development in Ravi Basin

Ravi has total catchment area of 5,451 sq. kilometers and 154 sq. km in Himachal Pradesh. This basin lies between the Pir Panjal and Dhauladhar ranges of the lower Himalayas. The potential capacity of Ravi has been assessed 2301.5 MW; out of which 1038.5 MW has been commissioned so far through four commissioned power projects in Chamba district, i.e. Baira Suil (198 MW), Chamera-I (540 MW), Chamera-II (300 MW) and excessively targeted (Table 5.4) for the hydroelectric power development since 1980s with the installation of NHPC's first hydroelectric power station (Baira Suil). After this power project Chamera, series of projects were started. Chamera-I[1] Power Station (3 × 180 MW) is a pondage scheme to harness the hydro power potential of river Ravi. It is located in Chamba district of Himachal Pradesh. The project comprises of a 121 m high, 295 m long concrete arch gravity dam with 9.5 in dia., 6.414 Km long head race tunnel. The underground power house has been installed capacity of 540 MW houses 3 units of 180 MW capacities each designed to operate under the net rated head of 185 m and designed to generate 1664.56 million units in a 90% dependable year with 95% machine availability . All the three generating units of the power station

Table 5.4: Showing Power Projects in Ravi Basin

Sr. No.	Name of Power Project	Name of River/Nallah	Classification	Installation Capacity (in MW)
1.	Chamera-I*	Ravi	Large	540.00
2.	Chamera-II *	Ravi	Large	300.00
3.	Chamera-III**	Ravi	Large	231.00
4.	Baira Suil *	Baira Suil	Large	198.00
5.	Bhuri Singh *	Saal Nala	Micro	000.45
6.	Tundah-I**	Tundah nallah	Small	015.00
7.	Tundah-II***	Tundah nallah	Small	030.00
8.	Bharmour Micro*	Budhil	Micro	000.02
9.	Garola**	Garola nallah	Micro	000.05
10.	Holi*	Holi nallah	Small	003.00
11.	Sal Stage-I***	Saal nallah	Small	008.25
12.	Sal Stage-II**	Saal nallah	Small	002.25
13.	Budhil**	Budhil nallah	Medium	081.00
14.	Bharmour**	Budhil nallah	Medium	045.00
15.	Harsar***	Budhil nallah	Medium	060.00
16.	Kugti***	Budhil nallah	Medium	045.00
17.	Kutehr**	Ravi	Large	360.00
18.	Sindi ***	Ravi	Medium	120.00
19.	Bara Bangal***	Ravi	Medium	160.00
20.	Thein Dam****	Ravi	Large	600.00

* indicates commissioned power projects.

** indicates under construction power projects.

*** indicates approved but yet to be started.

**** indicates power project situated at interstate boundary of HP, J&K and Punjab.

were commissioned in the month of April-1994. Chamera-II Power Station (3 × 100 MW) is a run of the river scheme with small pondage to harness the hydro power potential of river Ravi. It is located in Chamba district of Himachal Pradesh. The project comprises of a 43 m high, 118.5 m long concrete gravity dam with 7.0 m dia. 7.86 Km long head race tunnel. The underground power house with installed capacity of 300 MW houses 3 units of 100 MW capacities each designed to operate under the net rated head of 243 m and designed to generate 1499.89 million units in a 90% dependable year with 95% machine availability. Unit I, II and III of the power station were commissioned in the month Oct-2003, Dec-2003 and Feb-2004 respectively. Chamera-III[3] Power Station (3× 77 MW) is a run of the river scheme with small pondage to harness the hydro power potential of river Ravi. It is located in Chamba district of Himachal Pradesh. The project comprises of a

64 m high, 73 m long concrete gravity dam with 6.5 m dia., 15.97 Km long head race tunnel. The underground power house with installed capacity of 231 MW houses 3 units of 77 MW capacities each designed to operate under the net rated head of 200 m and designed to generate 1086.37/1108.17 million units in a 90% dependable year with 95% machine availability.

All the three generating units of the power station were commissioned in the month of June-2012. The beneficiary states/ UTs of this power station are Uttarakhand, UP, Delhi, HP, Haryana, J&K, Punjab, Rajasthan and Chandigarh. With the construction of the project, the area has also been benefitted by development of infrastructure, education, medical facilities and employment avenues.

The commissioned power projects in Ravi Basin, i.e. Baira-Suil, Chamera-I, Chamera-II and Thein dam of different magnitudes have intentionally altered the climatic conditions of the area. Particularly, the reservoirs of Thein dam (87sq. km) and chamera-I (29 km) have put pressure on the climatic conditions of the area and affected the agricultural practices.

About the Study

The present chapter is based on the study conducted by the researcher under UGC sponsored research project in Ravi basin in Chamba district of Himachal Pradesh. The prime objective of the study was to see the impact of dams on the lives, climate and sources of livelihood (agriculture) at micro level and how the impacts received are affecting the lives/livelihood of the people in the Ravi basin.

To achieve the broad objectives of the study, commissioned power project, i.e. Chamera-I in Chamba district of Himachal Pradesh, India was chosen. By applying exploratory and descriptive research method, using qualitative and quantative data, three hypotheses were constructed to be more empirical and specific, i.e. first hypothesis was framed to see whether the affected and native people are satisfied with the installation of power project and the compensation package given to them by NHPC; the second hypothesis was constructed to investigate the fact, the displaced people are happy because they have better living conditions as compared to the earlier and received their dues easily and well in time from NHPC; and last hypothesis was constructed to see the impact of power project on the social, economic as well as cultural status of the affected people has been raised with the installation of the power project.

For the scientific investigation, 22 villages out of 55 villages with different panchayats, developmental blocks, and administrative sub-divisions with the headquarters at Chamba were selected for data collection purpose. Keeping in view the geo-morphological consideration, 22 selected village(s) were divided in five research clusters. The respondents from 22 villages were divided in two categories, i.e. one those who have been affected and got displaced and living in some other villages and the second category of the respondents were of those who have been either partially affected or living in the vicinity of reservoir, HRT, TRT, power house or edits tunnels and may be called as the "Real bearer of the cost" of this developmental activity.

Being an exploratory and descriptive study, it was based on secondary as well as primary data. The first part of the study was based on secondary data available and the second part of the study was based on primary data collected specifically for the study. In the first part of the study, by reviewing available required information, base for the study was prepared by identifying variables; on the bases of which, second part of the study has been planned. In the second part of the study, micro level primary data based study was conducted on three identified objectives and three main hypotheses of the study. Further it is described in four chapters including the research methodology.

Impact of Dam on Climatic Conditions of the Study Area

On the basis of various studies conducted worldwide by earlier scholars and on the basis of research conducted by the researcher himself (2007 and 2009), under UGC sponsored research projects, before starting the actual study, it was assumed that dam catchment area is experiencing fluctuation in temperature and change in the wet days/rainy days. To see the impact of the dam installation on the climatic conditions of the area concerned, the data for 20 years; 10 years prior to the installation and 10 years after the installation of power projects in Ravi Basin was analysed.

Table 5.5: Amount of Rainfall in Millilitres (ml.) during the year 1996-2006 in Chamera-I Catchment Area

Year	Item*	Month												
		Jan	Feb	Mar	Apr	Mar	Jun	Jul	Aug	Sep	Oct	Nov	Dec	Total
1996	1*	121	--	168	48	71	95	158	290	103	--	---	---	1054*
	2**	04	--	08	02	04	08	13	11	04	--	---	---	54**
	3***	63	--	60	29	22	20	42	87	50	--	---	---	373***
1997	1*	09	79	17	103	76	175	152	132	85	70	90	115	1103*
	2**	03	04	02	07	05	16	07	09	07	09	04	03	76**
	3***	05	41	15	28	29	34	38	29	25	20	40	40	344***
1998	1*	35	245	130	55	15	121	60	85	165	53	--	40	1004*
	2**	02	09	06	04	02	03	04	08	07	07	--	08	60**
	3***	20	70	40	30	10	80	20	20	60	22	--	180	552***
1999	1*	180	25	25	15	15	20	216	135	95	--	100	--	826*
	2**	14	02	03	02	02	02	07	06	04	--	02	--	44**
	3***	200	20	10	10	10	10	60	40	40	--	60	--	460***
2000	1*	240	469	192	78	47	190	394	96	--	--	--	--	1706*
	2**	8	11	05	04	03	07	09	09	--	--	--	--	56**
	3***	60	67	60	38	25	55	75	45	--	--	--	--	425***
2001	1*	100	40	162	103	31	155	405	135	42	1	60	41	1275*
	2**	02	03	08	12	11	13	21	11	4	1	04	3	93**
	3**	75	20	65	31	11	29	98	60	4	1	48	20	462***
2002	1*	107	166	158	131	9	25	90	317	142	--	25	17	1187*
	2**	08	09	07	08	03	05	09	15	12	--	02	03	81**
	3***	40	40	28	27	05	10	30	50	40	--	15	15	300***
2003	1*	42	317	191	98	52	56	150	311	19	--	29	33	1298*
	2**	04	10	14	09	05	07	11	16	09	--	02	5	92**
	3***	36	80	40	38	28	16	54	16	80	--	17	15	420***
2004	1*	154	121	--	128	66	144	108	191	33	139	29	44	1157*
	2**	6	05	--	06	03	09	10	10	05	06	01	03	64**
	3***	60	48	--	45	50	44	36	63	08	56	29	26	465***
2005	1*	71	284	200	30	42	46	354	47	100	--	--	--	1174*
	2**	70	12	11	04	11	5	18	07	11	--	--	--	149**
	3***	12	50	67	13	7	17	45	13	23	--	--	--	247***
2006	1*	202	125	174	33	36	89	122	305	66	16	16	61	1245*
	2**	7	5	2	4	8	8	6	19	05	03	05	05	77**
	3***	68	52	27	28	17	31	26	40	37	6	08	29	369***

Source: Author's Compilation from the Data of the Meteorological Department of NHPC.

* By item in the table means (number-wise)

* Total amount of rainfall in a year (in ml)

** Number of days of rainfall in a year

*** Maximum rainfall during the day in a year (in ml)

Table

Amount of yearly total rainfall in millilitres (ml.) during the year 1996-2006 in Chamera-I Catchment Area

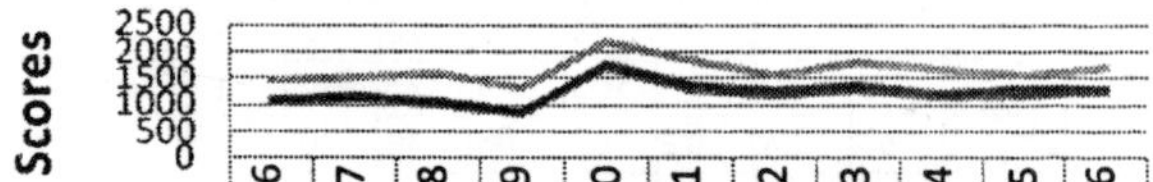

	1996	1997	1998	1999	2000	2001	2002	2003	2004	2005	2006
Maximum rainfall during the day in a year	373	344	552	460	425	462	300	420	465	247	369
Number of days of rain in a year	54	76	60	44	56	93	81	92	64	149	77
Total amount of rainfall in a year	1054	1103	1004	826	1706	1275	1187	1298	1157	1174	1245

On the basis of Table 5.5, it can be stated that after the installation of power project the area has experienced a lot of fluctuation in the climatic conditions. The total amount of rainfall in a particular year has experienced fluctuations as in 1996 it was 1054 millilitres (ml), goes upward in 1997 to 1103 ml while again in 1998 it goes up to 1004 ml while in 1999 it goes down to 826 ml and again in 2000 it arose to 1706 ml, in 2001 it again goes down to 1275 ml and remain continuously down in the next year to 1187 ml. The area has experienced 1298 ml rainfall in the year 2003 and in the next year it goes down to 1157 ml. The rainfall has gained upward mobility from 1157 ml to 1174 ml in the year 2005 and remained continue and touched 1245 ml point in 2006 (See Table 5.5 and Fig. 5.1).

On the basis of above description, it can be stated that during the course of past ten years the climatic conditions of the area had experienced a lot of weather jolts. Along with this, it is also experienced that the whole area has undergone many geological and meteorological changes. The area experienced many minor earthquakes jolts and newly erupted phenomena of cloud burst and unusual rain in this span of ten years, these all are the signs of climate change, which has increased after the installation of the power projects.

Average temperature from the period 1985 to 2002 as shown in the Table 5.7 shows that every year starting from 1985 to 2002

Table 5.6: Comparative Analysis of Monthly Average Values in Each Year for Wet Day Frequency in Chamba District of Himachal Pradesh Prior and After the Installation of Power Projects [Period: From 1986 to 2002] [Units: Integer, days]

Month	*Before the installation of HEPs from 1986-1993*	*After the installation of HEPs from 1994-2002*
January	29.09	37.77
February	35.66	41.90
March	43.36	33.67
April	29.09	36.25
May	33.31	35.60
June	44.29	51.52
July	85.00	87.93
August	77.89	106.93
September	47.34	43.07
October	11.99	14.69
November	11.03	12.20
December	16.15	14.37
Total	**464.20**	**515.90**

Source: Author's compilation from Tyndall Centre for Climate Change Research, Norwich, UK and *www.indiawaterportal.com*

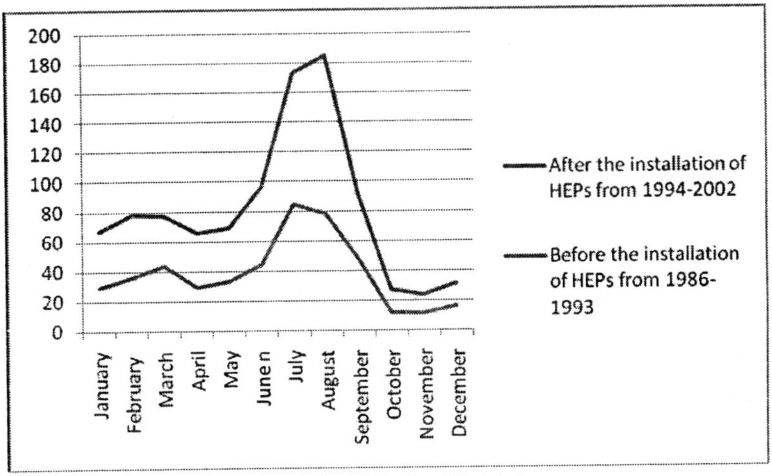

Fig. 5.1: Wet-Day Frequency in Chamba District.

has experienced upward increase to a considerable extent. The data of nine years prior to the installation of power projects is of fluctuating nature. It received fluctuations in every year whereas after the installation of power projects, average temperature has increased and from 1994 to 2002 it is fluctuating every month.

Table 5.7: Yearly Average Values in Each Year for Average Temperature Period: From 1985 to 2002 Units: Integer, degree Celsius

Prior to the installation of Power Projects		*After the installation of Power Projects*	
Year	*Average Temperature*	*Year*	*Average Temperature*
1985	237.74	1994	235.70
1986	223.64	1995	230.05
1987	232.04	1996	229.18
1988	238.34	1997	230.00
1989	225.14	1998	235.34
1990	235.26	1999	247.61
1991	226.85	2000	242.99
1992	229.73	2001	244.19
1993	233.23	2002	242.70
Total	**2081.97**	**Total**	**2137.76**

Source: Author's compilation from Tyndall Centre for Climate Change Research, Norwich, UK and *www.indiawaterportal.com*

On the basis of above presentation of data it can be stated that there is a very huge difference between the data prior to the installation and after the installation of power projects in Ravi basin. Because of these fluctuations, the agriculture, which has been the backbone of Indian society, has got a considerable jolt in the name of so-called development. It can be development for the nation or for the policy makers, but for the local people, who are bearing all these changes are compelled to receive less production in the fields and compelled to lead their life in misery and now searching other options of livelihood. Table 5.8 gives the detailed description of the production of crops by the people who are living in the vicinity of Chamera-I Hydroelectric Power Project in Ravi basin.

Table 5.8 is a detailed description of crop produced by the respondents in both conditions, i.e. before the installation of Chamera-I and at the time of interview, which presents a comparative analysis of both the conditions. Mostly crops like maize, wheat, rice, pulses, vegetables, etc. are being grown by the respondents. In the Table 5.8, quantity produced by the respondents and income generated out of the crop produced, suggested that with the installation of Chamera-I, the yielding of the crop and income generated out of crop has been decreased. There is in fact a very big difference in both the conditions.

Relationship of land with the farmers is not confined up to

the ploughing of the farms, but a complete way of living which broadly can be defined as agriculture. *Agri-* means farms or a place where farming takes place and *culture* means—total way of living, meaning thereby *it is a total way of living which revolves around the farms/cultivation*. Possession of land has a very vital role to play in the village economy as well as in the lives of the farmers.

Table 5.9 describes the status of possession of land by the respondents. Three sample variables including three types of land i.e. agricultural, homestead and pasture have been taken into consideration and both conditions (prior and after the installation of the power project) have been analysed. Data in the table suggests that land limit in *bighas* have decreased with the installation of the power project. As it is evident from Table 5.9 land possessed by the respondents in all three categories varies from 1 *bigha* of land to more than 50 *bighas*. There was not a single respondent who has responded that there was no land before the installation of the power project but after the installation of Chamera-I this number has increased to 90 respondents. The decreasing trend of the data shows that the installation of the power project has a detrimental relation with the decreasing holding and which further proves a trauma for the respondents. The remaining land has also received a negative impact because of a change in the climate of the villages situated in the vicinity of reservoir of Chamera-I which is responsible for reduction in the yield of the crops and lesser livelihood sustaining possibilities.

Conclusion

As it is evident from description that power projects in Ravi basin in Chamba district of Himachal Pradesh have given jolts to the physical environment by creating reservoirs, tunnels, edits, widening and construction work, which led the catchment area to get dried and the perennial Ravi is now seen either in reservoirs or in the tunnels. 154 kms long Ravi river is producing 1830 MW electricity and 45 other projects of different magnitudes in this basin are in different stages of their execution. This much disturbance has created a considerable effect on the climatic conditions of the whole area. The people who are living in the vicinity of these power projects are paying the cost, while some by living in metro cities are enjoying its fruits. Because of these climatic changes people are

Table 5.8: Detail of Crops Produced by the Respondents

Type of Crop(s)	*In OLD Village Before the installation of HEP*				*In NEW Village After the installation of HEP*			
	Yielding in quintal(s)	*Frequency*	*Income (in Rupees)*	*Frequency (percentage)*	*Yielding In quintal(s)*	*Frequency (percentage)*	*Income (in Rupees)*	*Frequency (percentage)*
Rice	Up to 2	27	Up to 2000	41	Up to 2	04	Up to 2000	04
	2-4	18	2001-4000	11	2-4	–	2001-4000	02
	4-6	09	4001-6000	06	4-6	02	4001-6000	–
	6-8	03	6001-8000	–	6-8	–	6001-8000	–
	8-10	01	8001-10000	02	8-10	–	8001-10000	–
	10+	02	100000+	–	10 +	–	100000+	–
	Total	**60***	**Total**	**60***	**Total**	**06***	**Total**	**06***
Wheat	Up to 2	23	Up to 2000	68	Up to 2	59	Up to 2000	78
	2-4	47	2001-4000	59	2-4	22	2001-4000	08
	4-6	52	4001-6000	38	4-6	06	4001-6000	01
	6-8	22	6001-8000	08	6-8	–	6001-8000	–
	8-10	17	8001-10000	04	8-10	–	8001-10000	–
	10+	26	100000+	10	10+	01	100000+	01
	Total	**187***	**Total**	**187***	**Total**	**88***	**Total**	**88***
Maize	Up to 2	15	Up to 2000	54	Up to 2	56	Up to 2000	75
	2-4	52	2001-4000	65	2-4	22	2001-4000	11
	4-6	32	4001-6000	29	4-6	07	4001-6000	–
	6-8	28	6001-8000	11	6-8	–	6001-8000	–
	8-10	15	8001-10000	05	8-10	01	8001-10000	–
	10+	34	100000+	12	10+	01	100000+	01

(*Contd.*)

	Total	176*	Total	176*	Total	87*	Total	87*
Pulse	Up to 20 kgs	37	Up to 200	15	Up to 20 kgs	03	Up to 200	–
	21-40	12	201-400	20	21-40	03	201-400	02
	41-60	13	401-600	09	41-60	–	401-600	03
	61-80	01	601-800	05	61-80	–	601-800	–
	81-100	–	801-1000	01	81-100	–	801-1000	–
	100+	02	1000+	15	100+	01	1000+	02
	Total	**65***	**Total**	**65***	**Total**	**07***	**Total**	**07***
Vegetables	Up to 10 kgs	07	Up to 100	07	Up to 10 kgs	01	Up to 100	01
	11-20	05	101-200	06	11-20	03	101-200	03
	21-30	08	201-300	07	21-30	–	201-300	–
	31-40	05	301-400	06	31-40	–	301-400	–
	41-50	05	401-500	05	41-50	–	401-500	–
	51+	08	501+	07	51+	–	501+	–
	Total	**38***	**Total**	**38***	**Total**	**03***	**Total**	**03***

*N is less than 200 because of multiple reasons.

Source: Field Survey.

Table 5.9: Detail of Possession of Land by the Respondents

Type of Possession	*Sample Variable*	*Sample Component*	*Total Number Possessed (in number)*			
			Before HEP (in OLD village)		*After HEP (in NEW village)*	
			Frequency	*Percentage*	*Frequency*	*Percentage*
1	2	3	4	5	6	7
Land (in bighas)	Agricultural	Up to 5	098	49.00	94	47.00
		5-10	059	29.50	10	05.00
		10-15	017	08.50	02	01.00
		15-20	016	08.00	01	00.50
		20-25	002	01.00	01	00.50
		25+	008	04.00	02	01.00
		No Land	--	--	90	45.00
		Total	**200**	**100.00**	**200**	**100.00**
	Homestead	Up to 1	91	56.80	82	65.60
		1-2	21	13.22	12	09.60
		2-3	08	05.00	03	02.40
		3-4	--	--	--	---
		4-5	01	00.62	01	00.80
		5+	04	02.50	--	--
		No Land	35	21.96	27	21.60
		Total	160*	100.00	125*	100.00

(*Contd.*)

1	2	3	4	5	6	7
	Pasture	Up to 10	23	76.66	04	18.18
		10-20	06	20.34	01	04.54
		20-30	01	03.00	--	---
		30-40	--	---	--	---
		40-50	--	---	--	---
		50+	--	---	17	77.28
		No Land		---	---	--
		Total	**30***	**100.00**	**22***	**100.00**

*N is less/more than 200 because of multiple reasons.

facing all sort of ill-effects like decrease in crop production, water mills are getting dried, static water of reservoirs is leading to health hazards and spreading of water-borne diseases, sources of livelihood have been snatched, etc. and the life of the people have become worse because of these power projects. It might be better for the people who are living far away and enjoying the electricity but the concerns of displaced people are waiting for the response from the executing agencies and also the concerns of the local people, who are living in the vicinity of these developmental mills remained unlistened and unsolved even after 15 years of Chamera-I execution and 46 years of commissioning of the two mega projects of Himachal Pradesh, i.e. Bhakra and Pong dams.

Undoubtedly, the targeted developmental goals have been achieved by the installation of these three mega projects, on three main rivers of the state, i.e. Bhakra on Satluj; Pong on Beas and Chamera-I on Ravi as national mission of river linking of three major rivers of Himachal Pradesh and watered the downstream states like Punjab, Haryana, Rajasthan etc. and brings prosperity in the northern part of the country. But the fact remains, the affected people of these two projects are still waiting for the justice and the people who remained there and living in the vicinity of reservoirs of these projects are the worse affected people in context of climate change.

In short, it can be asserted on the basis of above citations that not only the affected people in the revenue record but all the residents who are living in the Ravi basin in Chamba district are facing the ill-effects of these power projects in form of changes in the climatic conditions. People of the whole basin have never experienced the problems of increase in temperature or unusual and unseasonal rain before the installation of power projects but after the installation of power project (after 1994) it became a routine phenomena.

The Ravi catchment area falls in seismic zone V. The geological composition of this area comprises of rocks and other components which are in the formation stage. Such types of hydel electric power developments in the geologically weak zone are not desirable. As accepted by NHPC, the area of Baira Suil, Chamera-I and Chamera-II are geologically weak. After the installation of power projects in this area, the possibility of earthquake has increased manifolds. The World Commission on Dams (2000) in its report has also expressed the possibility of earthquakes measuring 3 to 3.5 Richter scale in the reservoirs as a routine phenomenon. In 2005, after the commissioning of Chamera-II, a number of mild earthquake jolts of 3.7 to 4.8 Richter scale have been experienced in Chamba. The earthquake of 1995 measuring 5.4 Richter scale immediately after the commissioning of Chamera-I in 1994 was the first retaliation of the nature with its epicentre at Kiani village, which is situated at the endpoint of 19-kilometer (Ravi side) reservoir of the Chamera-I.

Along with earthquakes, the whole area has also experienced the newly erupted phenomena of cloudburst. The cloudbursts are occurring almost every year and are responsible for the loss of property of the people and the fertile soil, trees and other forest commodities. So, it is submitted with conviction that such types of developmental activities are not desirable as well as viable especially in Seismic Zone V. The Ravi basin is being harnessed very badly from Madhopur to Bharmour (Table 5.1). Now the policy makers of State as well as at the national level must think judiciously before granting permission to instal such hydroelectric power projects.

The researcher is not against the installation of power projects but these must be eco-friendly and sustainable in nature and there must be sustainable improvement of human welfare. This means a significant advancement of human development, which is economically viable, socially equitable and environmentally sustainable. If the dams are the best way to achieve this goal, then this aspect deserves the full support of the researcher. The local issues must be taken into consideration properly and with true spirit, not in paper and must not be engulfed by the red tapism of the Indian bureaucracy. The policies should be framed by visiting local sites and taking local issues into consideration.

In conclusion, it can be said that let us commit ourselves to the creation of a sustainable 21st century that will be the inheritance of our children who will embrace and continue to manage water with gratitude for the coming generations.

NOTES

1. Chamera I: NHPC Limited, Website: nhpcindia.com/projectdetail.htm?catld=1&projected=11
2. Chamera II : nhpcindia.com/projectdetail.htm?catld=1&project 1d = 12
3. Chamera III : nhpcindia.com/projectdetail.htm?catld=1&project 1d = 30

6

The Environmental Impact of Unsustainable Food Practices

A Perspective on Food and Ecological Disaster

Radhika Mittal

Introduction

Man exists within nature and is in constant interaction with it. Ideally, human life mechanisms are built upon, in harmony with, and are interdependent with the environment. Primordial man may have been mystified and frightened by the elemental forces, but he also partook of natural abundance. He slowly grew to transform himself and his environment and may today be lulled into believing that he has mastery over it. Spirkin (1983) points out that man and nature interact dialectically—in a manner where man progressively believes that he depends less and less on nature, while indirectly, he becomes more and more dependent on it. Some human interference with the environment is for the purpose of sustenance, some for material and industrial development, and some for ambitious goals of comfort and luxury. The environment is a vast reservoir of resources and raw materials, but it is finite. Environmental sustainability has been crisply defined as the "maintenance of natural capital" (Goodland, 1995, p. 10) wherein, members and elements of the biophysical ecosystem that stimulate and sustain human life processes, are preserved, renewed, replenished and allowed to grow. The author aptly connects the social and economic sustainability with environmental sustainability, pointing out that social and economic processes today are contingent on inherited resources and finite capital.

Life indeed depends on life and human life processes rely on environmental systems and the non-human species that are a part of it. However, a delay and failure to acknowledge the current environmental harm will affect intergenerational sustainability (Goodland, 1995), leading to a catastrophic end, for nature will self-regulate herself, irrespective of man's follies. The author here warns against assuming that natural capital can be adequately substituted by man-made capital. The consumption of resources and extinction of species, he emphasizes, cannot be justified as an investment for more energy production or technological advancement. It would seem natural to understand that the oceans, breathable air, the topsoil and drinking water are not replenishable resources, but man has slowly established himself as a predator where nature is concerned.

Linking Food and Climate Change

Environmental sustainability requires an alert maintenance of natural capital and an equitable distribution of environmental wealth, along with an understanding that economic and social growth cannot be at the cost of extinction of resources and species. Some researchers (Dovers & Handmer, 1993) have called for humility, in place of arrogance, to acknowledge the destruction brought about by overconsumptive practices as a starting point to prepare for environmental sustainability.

In some ways, arrogance falls like a shadow over the discussion surrounding anthropogenic global warming or climate change. Despite clear scientific evidence and warnings, governments and corporations refuse to formulate and implement the necessary policies. Ruddiman (2003) suggests that anthropogenic environmental changes started as far back as 8000 years ago, when human settlements in Eurasia began agriculture and forest clearing. While it is not being suggested that humans turn the clock back and revert to becoming gatherers, the scale of biophysical deterioration wrought by overexploitation of the natural environment, does require some introspection, analysis and firm decision-making.

It is essential to identify the activities that contribute most significantly to global warming. It is often assumed that the combustion of fossil fuels for energy generation is the main source of human-induced greenhouse gas (GHG) emissions. While the contribution of fossil fuels to climate change cannot be neglected, scientists have recently started to draw attention to human food production and consumption practices. In the context of climate change, food production, which will be affected by widespread climatic changes in the near future, is a major contributor to those very changes through greenhouse gas (GHG) emissions. Climate change projections expect, besides extreme weather events, rising sea-levels and widespread pestilence and disease, extreme fluctuations in rain-fed agricultural output (IPCC, 2007).

Human food production and consumption practices have undergone a vast variety of dietary and technological changes since the second agricultural revolution. Bolstered by trade, economic growth, urbanisation and the Green Revolution, the capacity for food production and food choice has expanded. The most relevant changes include intensive animal farming, mechanisation and development of fertilisers and pesticides, genetic engineering and the cross-continental exchange and adaptation of crops to local soils and diets. This has led to the use of energy intensive practices, so much so, that in some countries, energy invested in food production outstrips the energy gained from food yield (McMichael, Powels, Butler & Uauy, 2007). Along with the need for a call to use renewable sources of fuel for energy production for all human activities, researchers across the world are sending out warning signals regarding unsustainable agricultural activities, in particular livestock production.

Animal Industries

Within the agricultural sector, animal farming accounts for a maximum of the sector's emissions—80 per cent by one estimation (McMichael *et al.*, 2007). The most recent studies (Goodland & Anhang, 2009, p. 11) go as far as to point out that livestock and their by-products are responsible for at least 51 per cent of annual, worldwide greenhouse gas emissions, a figure based on conservative estimates. The authors reveal that 25,048 million tons of carbon dioxide emissions, which should have been attributed to livestock, have been overlooked, misallocated or uncounted. They outline these categories in their report, from respiration by livestock, use of arable land for grazing livestock and growing feed to the calculation of methane emissions using an appropriate timeframe. The processes surrounding livestock production and consumption are equally unsustainable, from warming caused by fluorocarbons (required for cooling livestock products), to production, distribution and disposal of the by-products (leather, feathers, skin, fur), packaging materials and industry wastes (bone, fat and spoilages). Additionally, meat processing requires higher energy consumption and a greater use of fossil fuels.

In an attempt to offer a solution for meat-eating cultures to make the transition to a more sustainable diet, Goodland and Anhang (2009) recommend the use of meat analogs (made from **soy** and seitan) as well as dairy alternatives (rice, soy and oat milks and products). These products are already popular in some countries and a focused marketing approach by food companies, supermarkets, investment agencies and environmentalists would help create a strong business case. As economists, Goodland and Anhang (2009) imply that where arguments on health, environment and ethics may fail, the clarity of a workable business plan, advertising and consumer purchasing power might pave the path for a more sustainable future. **Soy** and wheat products, as well as brown rice and oats are naturally healthier than meat and dairy products and offer a strong nutrition profile (Erdman & Fordyce, 1989; Jacobs & Steffen, 2003), along with an environmental advantage (Reijnders & Soret, 2003). By slowly eliminating the need for livestock products, the land as well as the grain that is produced to feed farmed animals can be utilised for human consumption. This will also be a step towards freeing up land and water resources, curtailing waste generation, addressing world hunger and promoting employment through just labour practices.

To further support the need for a climate-friendly diet, Swedish researcher Carlsson-Kanyama (1998), in an analysis of GHG emissions from carrots, tomatoes, potatoes, pork, rice and dry peas, finds that a meal consisting of tomatoes, rice and pork in Sweden has nine times higher emissions than a meal comprising potatoes, carrots and dry peas. In analysing GHG emissions and energy consumption in the production processes of these foods, individually as well as in four meal variants comprising some or all of the six items, she finds evidence for the unsustainability of food derived from animals or imported from other countries (tomatoes and rice, in the case of Sweden). Although transportation would have been assumed to be a major contributor to the emissions, the author finds that the various stages of production, such as pig rearing, crop farming, storage and production of fertilizers affect the final calculations in a bigger way. Making a strong case for bringing about a change in the diets common in developed economies, Carlsson-Kanyama (1998, p. 292) clarifies that current consumption patterns exceed the level of sustainability required by at least a factor of 4.

In a similar study with a broader, more extensive range of food products, researchers (Carlsson-Kanyama & Gonzalez, 2009) found that fresh vegetables, cereals and legumes have the lowest GHG emissions, while meats (especially from ruminants) and dairy have much higher emissions. While emissions from fish are in the middle range, the severe depletion and overexploitation of fish stocks around the world (Jacquet, Pauly, Ainley *et al.*, 2010; FAO, 2005) does not augur well for the future of fish consumption. In effect, fruit, vegetables, cereals and legumes are the most sustainable food products (except for those transported by air, especially fruit). Land use for factory farming leads to long-term soil erosion with little or no time for recuperation. Nutrient loss from soils results in the application of energy-intensive chemical fertilizers, which in turn affect soil fertility through prolonged use (Pimentel, Acquay, Biltonen *et al.*, 1992). Animal farms are spread over almost one-third of the earth's land area and another third of the world's arable land is cultivated for animal feed (McMichael *et al.*, 2007). An increase in the demand for meat in developing economies has encroached upon tropical rainforests in South America and left farmers across the world with little or no land to cultivate crops on (Naylor, Steinfeld, Falcon *et al.*, 2005). Livestock farming is slowly, but surely, draining the Earth's capacity to sustain.

The Waste of Protein

Climate efficiency for producing protein from vegetable sources is much greater and Carlsson-Kanyama & Gonzalez (2009) point out, as an aside, that a variety of vegetable sources of protein have not been duly acknowledged. The story and cycle of protein is well understood in climate and health science, but misconceptions abound amongst the general public and seem to be perpetuated by policy makers. In an environmental context, rearing animals for food entails feeding them 6 kg of plant protein to produce 1 kg of superior animal protein (Pimentel & Pimentel, 2003). This 1 kg of animal protein utilizes about 100 times more water than would be needed to produce 1 kg of grain protein (Pimentel & Pimentel, 1996). Legumes like soya beans, for instance, are a great source of protein, and require a fraction of the energy needed to produce animal protein. Ironically, grains and legumes are produced for animal feed, rather than human beings, contributing to a vicious cycle of environmental decay, poverty, malnutrition and hunger. The US livestock population alone consumes more than 7 times the grain that the human American population consumes (USDA, 2001). This would be startling enough if it were not made more absurd by the fact that Americans today consume twice the amount of protein that they are recommended to by their own government agencies (Pimentel & Pimentel, 2003) and subsequently, will require 5.4 planets to sustain their growing needs (Wackernagel and Rees, 1996).

Does the extensive research generated on livestock industries absolve other food practices of their contribution to a deteriorating environment? No, it does not. However, it should be noted that apart from carbon dioxide emissions, nitrous oxide emissions (from fertilizer application, soil and manure management) and methane emissions (from enteric fermentation in ruminant animals and manure management) form a large component of GHG emissions from food (Weber & Matthews, 2008). These too are focused around the production of red meat and dairy. Associated food practices from food miles to conventional farming to genetic modification need to be taken stock of as well; but their contribution is on a different scale.

Food Miles

Food transported over long distances, especially by air, refrigerated food (of all types), and some horticultural practices do contribute to GHG emissions (Jungbluth, Tietje & Scholz, 2000). While it would be advisable to eat local to reduce this impact of 11 per cent, it has been found that shifting to a vegetable-based diet achieves more GHG reduction than purchasing all locally sourced food (Weber & Matthews, 2008). Eating locally has many other advantages in the form of providing support to local farmers and local economies. The increasing availability of local and organic food in the US and UK, therefore, does signal a positive trend. Stagl (2002) avers that local and organic markets offer a space to renew and replenish social-environment relationships. In this context, it is pertinent to ask whether the demand for exotic and branded food items, a possible developing trend in countries like India, is sustainable.

Agricultural Practices

Agricultural practices that would be considered more sustainable than conventional production include low-intensive and traditional farming, organic farming and integrated farming (Pretty, 1998). Amongst these, organic farming is the most efficient, reducing carbon dioxide emissions by 50 per cent and using 40 per cent less energy for the same output, in comparison to conventional farming (Haas & Köpke, 1994, as cited in Sundrum, 2001). Organic agriculture has a positive impact on the environment, from improving soil and water quality to supporting biodiversity and contributing to ecological harmony. The effects range from reducing soil erosion, improving the physical, chemical and biological make-up of the soil to promoting the growth of beneficial organisms naturally found in the flora and fauna (Conacher & Conacher, 1998; Liebig & Doran, 1999). Additionally, recent studies have found that organic agriculture is capable of producing enough yields to sustain the human population with its current calorie intake (Badgley & Perfecto, 2007).

Conventional agriculture, over time, with mechanised, chemically-driven farming practices has led to greater energy expenditure, waste generation and resource depletion. Pesticides and insecticides used in conventional agriculture are not only considered harmful for human health, but, in disturbing the natural ecological balance, also spawn a range of problems (Conway & Pretty, 1991; Pimentel, McLaughlin, Zepp *et al.*, 1991). They diminish soil nutrients; poison birds, fishes (Grist, 1986) and animals higher up in the food chain, create an uneven balance between predator and prey, particularly in the insect world (Brambel, 1989) and, at times, lead to outright crop destruction (Pimentel *et al.*, 1992).

Fertilizer and pesticide application are the most important secondary sources of GHG emissions, after tilling and inefficient irrigation (Lal, 2004). As suggested by the author, more energy efficient methods for tilling (such as seeding after chiselling), irrigation (such as furrow or trickle irrigation as opposed to flood irrigation) and fertilizer/pesticide usage (such as biological nitrogen fixation, recycling nutrients in crop residue) can be adopted. In some crops, the use of fertilizers is the greatest source of energy input (Pimentel, Hurd, Bellotti *et al.*, 1973). Of all the unsustainable agricultural practices, fertilizers and pesticides work at the very core of destroying natural resources by depleting natural soil fertility, leading to air and groundwater pollution and contamination (Cowling, Galloway, Furiness *et al.*, 2001) and leaving behind residues that find their way into food (Pimentel *et al.*, 1992). Moreover, farmers would find it most difficult to escape the entrapment created by a dominant culture of chemical usage and throw off the weight of pest control technology in a controlling market system bearing down on them in the face of economic survival (Wilson & Tisdell, 2000). This has implications beyond environmental sustainability, although those are severe enough.

Dubious and worrisome practices include using banned pesticides (Wilson, 1998) and mixing a variety of pesticides to create stronger resistance to pests (Chandrasekera, Wettasinghe & Amarasiri, 1995). Moreover, prolonged usage just gives rise to pest populations which are resistant to these concoctions, creating a situation where their application is no longer cost-effective (Wilson & Tisdel!, 2000). Pesticide usage has seen a growth over the past decades, both in developing economies and in per acre usage of land (WRI, 1998). Farmers from poorer countries are at a greater risk to the ill-effects of pesticide usage due to overexposure (such as hand-spraying), lack of protective gear, more toxic insecticides and inadequate training and education (Wilson & Tisdell, 2000).

The application of technology to generate greater agricultural productivity has become more intensive over the years. The issue that needs to be addressed is: what is the ecological cost to the world? Perhaps the answer can only be found when we step back and review the methods being used currently. The process of manipulation of genetic material through genetic engineering of plants or animals for human consumption is one such method.

Genetic modification (GM) has wide-ranging implications for the environment and human health and the risk of allowing severe drawbacks to surface over an extended period of time may not be worth the risk. Environmental risks, as outlined by Pretty (2001) include, the alteration of native plant ecology due to the transfer of genes, increased, new forms of insect and weed resistance, secondary pest problems such as GM crops becoming problem weeds themselves, the development of new types of pathogens due to recombining transgenes and the development and effects of new toxins expressed by GMOs. The author points out that farmers using GM crops use more herbicides (broad-spectrum) which are capable of wiping out weeds completely. At the same time, these stunt the growth of all herbivorous insects and other seeds (except for the GM crop), leading to a disturbance in the food and life cycle of birds, mammals and other organisms that comprise the ecosystem. Where human health is concerned, the risks are of developing newer allergic and immune system reactions (Pretty, 2001; Anon, 1999) as well as antibiotic resistance (Grassen, 2000). More exhaustive claims have been levelled at the use of such a technology, and not one without reason. GMO concerns include preservation of seed diversity, seed access (Shiva, 1996), local and organic farmers' rights (Koch, 1998), religious, cultural and ethical questions (Robinson, 1997; Crist, 1996) and labelling, or the lack of it. It is beyond the scope of this chapter to discuss each potential risk within the practices employed today for food production, but it is necessary to open a window into the practices of consumption.

Need of the Hour

With a population hurtling towards 7 billion people, around 2 billion consume a meat-based diet and around 4 billion a plant-based one. More than 3 billion people are malnourished according to the World Health Organisation (WHO, 2000, as cited in Pimentel & Pimentel, 2003), but the 2 billion that flourish on a meat-based diet, to satisfy their tastes, desires or out of pure habit, continue to consume a large chunk of the environmental resources that sustain the earth (White, 2000). This is an inequitable scale on which we exist and future projections, through the mirror of aspirations, are dark and gloomy.

The trend in developing economies of the south is to emulate the north. With urbanisation, increasing incomes and fast-paced lifestyles, households across India, China, Latin America and Sub-Saharan Africa, are seeking energy-dense foods, rich in sugars and animal fats. Delgado (2003) points out that China will lead the way, with Brazil a close second, in almost doubling the total quantity of meat consumed by 2020. In the same projection, India and other South Asian countries will drive an enormous increase in the consumption of milk products. In the past 40 years, the nations of the South have transitioned to consuming a diet high in vegetable oils, meat, seafood and sugar, away from pulses, legumes, roots and tubers (Kearney, 2010). Ironically, it has been pointed out, in the same study, that worldwide consumption of fruits and vegetables is inadequate, despite an increase in production.

Global practices of food consumption are unsustainable, both at an ecological level and from a medical/health perspective. Meat and milk consumption has been directly associated with a rise in obesity, cardiovascular diseases and cancers (Barnard, Cohen, Jenkins *et al.*, 2009; Craig, 2009; Fraser, 2009; Lanou, 2009), escalating further health care costs and necessitating energy-intensive interventions. The future of human health appears to be one of overnutrition and disease epidemics (Kearney, 2010). Consumers seem to be motivated by new definitions governing social status, aspirations and expectations. It has been argued that man is living in Anthropocene—an artificial ecology of his own making (Dalby, 2009). If that is indeed the case, then probably no amount of information and right direction will be able to stir him out of his reverie. However, is there an integrated effort at providing this direction?

The questions that need to be addressed and have been hinted at through the research outlined in this chapter are:

1. Why does climate policy, across governments, not reflect the benefits of plant-based diets for the environment, health and socio-economic well-being for the world at large?
2. Why is the clear evidence and information about the unsustainability of food production and consumption practices not easily accessible by consumers through consumer forums, climate activists, health educationists and the news media?

It is important to analyse and amend the entire structure of food production, distribution and consumption systems. No single process can be targeted as more detrimental when it exists within a broader framework of consistent unsustainability. For example, shopping in a cooperative store, as opposed to a supermarket reduces wastage by 75 per cent (in energy, air and water pollution) according to a Viennese Study (Erzeuger-Verbraucher-Initiative, 1992, as explained by Stagl, 2002). This example suggests that changes need to be made at a broader social level. Human beings exist within the environment and their ways of living and social structures cannot be made in isolation, divorced from the reality of their existence. While one of the most important decisions may be to move away from livestock farming, displaying a true reverence and respect for the creatures that we share our world with, it must be accompanied by creating a more equitable, compassionate and just socio-economic system.

The Indian Perspective

While across the North, recommendations are being made to transition to a more sustainable food system through improved practices and dietary choices (Reijnders & Soret, 2003), the South, including India, is aspiring to repeat the mistakes of the North. This is a sharp deviation from the land where the concept of "*vasudhaiva kutumbakam*, i.e. all that is alive, from plants to human species, belongs to a single family" (Misra, 1997) was born.

The Vedic texts of India have venerated nature and maintained that the environment has a moral order essential to human survival. Elements of nature, like the Himalayan mountain range, rivers like the Ganga and trees like the Sacred Fig (*Ficus religiosa* in Latin or *Ashvattha* in Sanskrit) have been imbued with elements of divinity to inspire the acceptance of natural surroundings as creative and protective energies (Jha & Burdhan, n.d.). The concepts of divinity have been intertwined with nature, with the acceptance and protection of all species and elements, creating an everyday, lived spirituality. The *Isha-Upanishads* cautioned against any kind of encroachment, facilitating an atmosphere of mutual respect and symbiosis (Ramasubramaniam, 2008). Not only was there a synergy in one's own behaviour, there were clear references to punishment for those who were abusive, exploitative and attempted to establish dominion over nature. Not so long ago, taking a leaf from India's ancient wisdom, Mahatma Gandhi wrote: "God forbid that India should ever take to industrialization after the manner of the West. The economic imperialism of a single tiny island kingdom (England) is today keeping the world in chains. If an entire nation of 300 million took to similar economic exploitation, it would strip the world bare like locusts" (Guha, 2006, p. 231).

In an ego-centric, materialistic, overconsumptive environment, the values of a civilisation that knew how to live in an environmentally sustainable manner seem to be lost. Their development had been enabled by a mental and social construct of harmony and by considering the self as a part of the ecology within which it is contained (Gandhi, 1884-1946). Although, there is a possibility of bringing this wisdom forth for a vision and reality of sustainable development, it needs to be practiced before it can be preached.

REFERENCES

Anon. (1999). "Editorial". *The Lancet,* 253 (9167): 1811.

Badgley, C., and Perfecto, I. (2007). "Can organic agriculture feed the world?" *Renewable Agriculture and Food Systems,* 22, 80-86, Cambridge University Press.

Barnard, N., Cohen, J., Jenkins, D. *et al.,* (2009). "A low-fat vegan diet and a conventional diabetes diet in the treatment of type 2 diabetes: a randomized, controlled, 74-wk clinical trial." *American Journal of Clinical Nutrition,* 89, 1588S-96S.

Bramble, B.J. (1989). "An environmentalist view of pest management and the Green Revolution." *Tropical Pest Management,* 35 (3), 228-230.

Carlsson-Kanyama, A. (1998). "Climate change and dietary choices—how can emissions of greenhouse gases from food consumption be reduced?" *Food Policy,* 23 (3/4), 277-293.

Carlsson-Kanyama, A. and Gonza'lez, A.D. (2009). "Potential contributions of food consumption patterns to climate change." *American Journal of Clinical Nutrition,* 89, 1704S-9S.

Chandrasekara, A.I., Wettasinghe, A., Amarasiri, S.L. (1985). *Pesticide usage by vegetable farmers.* Paper presented at Annual Research Conference ISTI, Gannoruwa, Sri Lanka.

Conacher, J., and Conacher, A. (1998). "Organic farming and the environment, with particular reference to Australia: A review." *Biological Agriculture and Horticulture,* 16 (2), 145-171.

Conway, G.R. and Pretty, J.N. (1991). "Unwelcome Harvest: Agriculture and Pollution." *Earthscan*, London.

Cowling, E., Galloway, J., Furiness, C. *et al*., (2001). "Optimizing Nitrogen management in food and energy production and environmental protection." Report from the Second International Nitrogen Conference, USA. Available from http://www.initrogen.org/fileadmin/user_upload/Second_N_Conf_ Report.pdf

Craig, W.J. (2009). "Health effects of vegan diets." *American Journal of Clinical Nutrition*, 89, 1627S-33S.

Crist, W.E. (1996). "Waiter, there's a flounder in my fruit. (Bio-engineered fruits and vegetables with animal genetic materials are not so labeled)." *Vegetarian Times*, 231(22).

Dalby, S. (2009). *Security and Environmental Change*. Cambridge: Polity.

Delgado, C. (2003). "Rising consumption of meat and milk in developing countries has created a new food revolution." *The Journal of Nutrition*, 133, 3907S-3910S.

Dovers, S.R. and Handmer, J.W. (1993). "Contradictions in Sustainability." *Environmental Conservation*, 20 (3), 217-222.

Erdman, J.W. and Fordyce, E.J. (1989). "Soy products and the human diet." *American Journal of Clinical Nutrition*, 49, 725-37.

FAO (2005). *The State of Food Insecurity in the World 2005*. Food and Agriculture Organization of the United Nations. Rome.

Fraser, G.E. (2009). "Vegetarian diets: what do we know of their effects on common chronic diseases?" *American Journal of Clinical Nutrition*, 89, 1607S-12S.

Gandhi, M.K. (1884-1946). *The Collected Works of Mahatma Gandhi*. New Delhi: Government of India, Ministry of Information and Broadcasting.

Gassen, H.G. (2000). "Possible risks of GM foods." Paper presented at OECD Edinburgh conference on *The Scientific and Health Aspects of GM Foods*, Available from http://www.oecd.org/subject/biotech/ed_prog_sum.htm

Goodland, R. (1995). "The Concept of Environmental Sustainability." *Annual Review of Ecology and Systematics*, 26, 1-24.

Goodland, R. and Anhang, J. (2009). "Livestock and Climate Change." *World Watch Magazine*. Available from http://www.worldwatch. org/node/6294

Grist, D.H. (1986). *Rice*. Sixth Edition. Longman, London.

Guha, R. (2006). *How much should a person consume: Environmentalism in India and the United States*. Berkeley: University of California Press.

IPCC. (2007). *Climate change 2007: Synthesis report*. Available from http://www.ipcc.ch/pdf/assessment-report/ar4/syr/ar4_syr_spm.pdf

Jacobs, D.R. and Steffen, L.M. (2003). "Nutrients, foods, and dietary patterns as exposures in research: a framework for food synergy." *American Journal of Clinical Nutrition*, 78, 508S-13S.

Jacquet, J., Pauly, D., Ainley, D., Holt. S., Dayton, P. and Jackson, J. (2010). "Seafood stewardship in crisis." *Nature*, 467, 28-29

Jha, S.K. and Burdhan, A. (n.d.) *Classical Indian Concept of Environmental Protection and Art*. Indira Gandhi National Centre for the Arts, Available at http://ignca.nic.in/sjha0003.htm

Jungbluth, N., Tietje, O., Scholz, R.W. (2000). "Food purchases: impacts from the consumers' point of view investigated with a modular LCA." *The International Journal of Life Cycle Assessment*, 5, 134-42.

Kearney, J. (2010). "Food consumption trends and drivers." Philosophical Transactions of The Royal Society, *Biological Sciences*, 365 (1554), 2793-2807.

Koch, K. (1998). "Food safety battle: organic *vs.* biotech." *Congressional Quarterly Researcher*, 9 (33), 761-84.

Lal, R. (2004). "Carbon emission from farm operations." *Environment International*, 30, 981-990.

Lanou, A.J. (2009). "Should dairy be recommended as part of a healthy vegetarian diet? Counterpoint." *American Journal of Clinical Nutrition*, 89, 1638S-42S.

Liebig, M.A., and Doran, J.W. (1999). "Impact of organic production practices on soil quality indicators." *Journal of Environmental Quality*, 28 (5), 1601-1609.

McMichael, A.J., Powles, J.W., Butler, C.D. and Uauy, R. (2007). "Food, livestock production, energy, climate change and health." *The Lancet*, 370 (9594), 1253-1263.

Misra, R.P. (1997). "The Indian world-view and environmental crisis." In B. Saraswati (Eds.), *Integration of endogenous cultural dimension into development*. Available from http://ignca.nic.in/cd_05016.htm

Naylor, R., Steinfeld, H., Falcon, W. *et al.*, (2005). "Losing the Links Between Livestock and Land. *Science*, 310, 1621-1622.

Pimentel, D., Acquay, H., Biltonen, M., *et al.*, (1992). "Environmental and economic costs of pesticide use." *Bioscience*, 42 (10), 750-760.

Pimentel, D., and Pimentel, M. (1996). *Food, energy and society*, Niwot, CO: Colorado University Press.

Pimentel, D., and Pimentel, M. (2003). "World population, food, natural resources, and survival." *World Futures*, 59, 145-67.

Pimentel, D., Hurd, L.E., Bellotti, A.C. *et al.*, (1973). "Food production and the energy crisis" *Science*, 182 (4111), 443-449.

Pimentel, D., McLaughlin, A., Zepp, B. *et al.*, (1991). "Environmental and economic impacts of reducing U.S. agricultural pesticide use." In D. Pimentel, (Ed.) *Handbook of Pest Management in Agriculture*. CRC Press, Boca Raton, FL.

Pretty, J. (1998). *The Living Land: Agriculture, Food and Community Regeneration in Rural Europe*. London, UK: Earthscan Publications Ltd.

Pretty, J. (2001). "The rapid emergence of genetic modification in world agriculture: contested risks and benefits." *Environmental Conservation* 28 (3), 248-262.

Ramasubramaniam, S. (2008). *Relevance of traditional Indian value system to ecocity*. Ecocity World Summit 2008 Proceedings, California.

Reijnders, L. and Soret, S. (2003). "Quantification of the environmental impact of different dietary protein choices." *American Journal of Clinical Nutrition*, 78, 664S-8S.

Robinson, C. (1997). "Genetically modified foods and consumer choice." *Trends in Food Science and Technology*, 8, 84-8.

Ruddiman, W.F. (2003). "The anthropogenic greenhouse era began thousands of years ago." *Climate Change*, 61, 261-293.

Shiva, V. (1996). "Agricultural biodiversity, intellectual property rights and farmers' rights." *Economic and Political Weekly*, 31 (25), 1621-1631.

Spirkin, A. (1983). Dialectical materialism, Available from http://www.marxists.org/reference/archive/spirkin/works/dialectical materialism/index.html\

Stagl, S. (2002). "Local organic food markets: potentials and limitations for contributing to sustainable development." *Empirica*, 29, 145-162.

Sundrum, A. (2001). "Organic livestock farming: a critical review." *Livestock Production Science*, 67 (3), 207-215.

USDA (2001). "Agricultural Statistics." Washington, DC: US Department of Agriculture.

Wackernagel, M., and Rees, W.E. (1996). *Our ecological footprint: reducing human impact on the earth*. Gabriola Press New Society Publishing, B.C.

Weber, C.L., and Matthews, H.S. (2008). "Food-Miles and the relative climate impacts of food choices in the United States." *Environmental Science and Technology*, 42 (10), 3508-3513.

Wilson, C. and Tisdell, C. (2000). "Why farmers continue to use pesticides despite environmental, health and sustainability costs." *Ecological Economics*, 39 (3), 446 462.

WRI (1998). *World Resources, 1998/1999*. World Resources Institute, Oxford University Press, UK.

7

Climate Change and Disasters

Uttar Pradesh in Context

K.N. Bhatt

Introduction

The climate can be thought of as a long-term summary of weather conditions taking account of the average conditions as well as the variability of these conditions. Both climate change and disaster risk reduction are closely linked as the existing methods and tools of disaster risk reduction (DRR) provide powerful capacities for adaptation to climate change.[1] In the past, the Earth's climate has varied considerably. However, global temperatures have now risen unusually rapidly. The concentration of greenhouse gases in the atmosphere is today the highest in the past 5,00,000 years. This has grown by 70 per cent between 1970 and 2004 alone. It is widely accepted that human activities are the dominant factors in this growth and are responsible for most of the warming observed over the past 50 years.[2]

The Intergovernmental Panel on Climate Change (IPCC) views climate change as "a change in the state of the climate that can be identified...by changes in the mean and/or the variability of its properties.... It refers to any change in climate over time, whether due to natural variability or as a result of human activity."[3] The United Nations Framework Convention on Climate Change (UNFCCC) defines climate change as a change of climate that is attributed directly or indirectly to human activity that alters the composition of the global atmosphere and that is in addition to natural climate variability observed over comparable time periods."[4]

Increased intensity of natural hazards as a result of climate changes turn into disasters when a combination of a vulnerable, exposed, and ill prepared population encounters a hazard event. Climate change therefore affects disaster risks in two ways: first by the likely increase in weather and climate hazards, and secondly through increases in the vulnerability of communities to natural hazards. Environmental degradation, unprecedented population growth and rapid unplanned urban expansion add yet another stress, further reducing abilities of people to cope with even the existing levels of weather hazards.[5] Frequent occurrences of disasters put development gains at risk and adversely impact human lives, damage ecosystems and strain livelihoods.[6]

The developing countries are affected more seriously by disasters as two-third of the victims belong to them.[7] During the period 1991-2005 about 3,470 million people were globally affected by disasters, 9,60,000 people died, and incurred economic losses amounting to US$ 1,193 billion. In last two decades 1988-2007, 76 per cent of all disaster events were hydrological, meteorological or climatological in nature. These events accounted for 45 per cent of the deaths and 79 per cent of the economic losses. The likelihood of increased weather extremes in future therefore poses great concern.[8]

Asia experienced the most devastating disasters of the world including the Tangshan earthquake of China in 1976 (causing 2,55,000 deaths), the Indian Ocean Tsunami, 2004 (resulting in 2,83,000 deaths), and the Bangladesh cyclone and floods in 1970 and 1991 (leading to 3,00,000 and 1,83,000 deaths). It is noted that countries of this region lose between 2 per cent to 20 per cent of their GDP and 10 per cent to 30 per cent of their revenue every year on account of disasters. Besides, a sizeable portion of their resources get diverted to disaster relief and rehabilitation. Disasters in the Asian continent are identified as one of the major factors that perpetuate the condition of poverty in many of its parts.[9]

This chapter analyzes the outcomes of disaster risk management projects implemented in Uttar Pradesh during 2004-2008. It documents some case studies/success stories from project covered districts to learn lessons for policy interventions in the future. It also examines the interlinkages between the incoming dangers of climate change and disasters. The Government of India (GoI)-UNDP initiated Disaster Risk Management Programme (DRMP) 2002-07 dovetailing all national programmes covered under the UNFCCC. A sub-component of DRMP focused on sustainable reduction in earthquake risk in the most earthquake prone urban areas across the country.

Disaster Management: An Overview

The UN General Assembly in 1989 declared the decade 1990-2000 as the International Decade for Natural Disaster Reduction.[10] The Asian city Yokohama hosted the First World Conference on Disaster Risk Reduction (DRR) in 1994 where the Yokohama Strategy for a Safer World was adopted. Japanese city of Kobe hosted the Second World Conference in 2005 which adopted the Hyogo Framework for Action (HFA) 2005-2015. The HFA adopted five priorities for action through five fold processes including: political, technical, socio-educational, developmental and humanitarian processes. All these processes are considered crucial for achieving MGDs (2000) and the Sustainable Development Plan (2001) of the United Nations.[11] In Asia, the Association of South East Asian Nations (ASEAN) adopted ASEAN Agreement on Disaster Management and Emergency Response. The South Asian Association for Regional Cooperation (SAARC) finalized a comprehensive Framework of Disaster Management, which was adopted at the Delhi Summit in April, 2007.[12]

India is highly vulnerable to recurring disasters like floods, droughts, cyclones, earthquakes and landslides. About 60 per cent of the country's landmass is reported prone to earthquakes; over 40 million hectares is prone to floods; almost 8 per cent is vulnerable to cyclones and 68 per cent area is susceptible to drought. As per a World Bank estimate, the disaster losses, during 1996-2001, including Orissa super cyclone in October 1999 and the Gujarat Bhuj earthquake in January, 2001, amounted to US $ 13.8 billion.[13]

India's Tenth Five Year Plan (2002-07), set into motion the process of shift in focus from response-centric disaster management to prevention, mitigation and preparedness. Subsequently, the Eleventh Plan aims at consolidating the process that develop and nurture the culture of safety and the integration of disaster prevention and mitigation into the development process. The guidance and direction to achieve this paradigm shift is expected to emanate from the National Disaster Management Authority (NDMA) and in the true spirit of the Disaster Management Act, 2005.[14]

Disaster Risk Management (DRM) is a multidisciplinary field covering a wide range of issues such as monitoring, evacuation, search, rescue, relief, reconstruction and rehabilitation. It also requires multi-sector governance involving scientists, planners, volunteers and communities. Their roles and activities span the pre-disaster, during the disaster and the post-disaster phases. Thus, the efficient functioning of disaster mitigation and management programmes in a country are important measures for its sustainable development, safety of human lives and protection of natural environment.

Disasters in Uttar Pradesh

Uttar Pradesh is prone to diverse kinds of disasters: floods, droughts, fires and earthquakes.[15] Out of the state's 240.93 lakh hectares geographic area, 73.06 lakh hectares is sensitive to floods, incurring a financial loss of Rs. 2,000 crores every year and causing destruction to its economy. Many major rivers of India flow from within the state's boundary in addition to numerous small rivers and streams. Floods are caused mainly by two factors: (i) heavy rainfall causing waterlogging and (ii) release of water from the barrages resulting in floods.[16]

The total area of 16 districts and parts of another 13 districts of U.P. fall under High Earthquake Damage Risk Recurring Zone-IV.[17] In rural areas often several incidents of fire engulf the *kutcha* houses with straw roofs. Fires often ruin standing crops or those stacked in granaries. Fires also occur in urban and industrial areas.[18] The recurrence period of highly deficient rainfall in Eastern U.P. resulting in drought is estimated 6-8 years, whereas in Western U.P. it is 10 years. In recent times, 2002 and 2004 were severe in terms of drought with an estimated loss of Rs. 7,540 crores and Rs. 7,292 crores respectively. Bundelkhand region is the most deficient in rainfall for the last 4 to 5 years. Like drought extreme cold and heat conditions, hailstorm and cloudbursts, also cause irreparable loss to life, crops and property.[19]

DRMP in Uttar Pradesh

In Uttar Pradesh, the 'Community Based Disaster Risk Reduction (CBDRR) through Participation of Communities and Local Self-

Government', UNDP-India, project has been implemented in 13 districts, while the Urban Earthquake Vulnerability Reduction (UEVR) project was run in 6 cities during 2004-09. The CBDRR project focused to reduce the impact of disasters, enhance management skills and improve capacities.[20] The UEVR aimed at strengthening capacity of communities, urban local bodies and the administration in mitigation, preparedness and response.[21]

Accordingly, all the programme districts have developed various types of capacities to all stakeholders in the areas of mitigation, search and rescue, first-aid, relief, and rehabilitation. The implementation phase of both the projects ended on December 31, 2008. In order to learn lessons from various activities conducted under these projects, Government of Uttar Pradesh and UNDP assigned a study to G.B. Pant Social Science Institute, Allahabad[22] to identify cases/success stories from the project districts, analyze and document them and highlight the lessons learnt for policy interventions in future.

Location Map of UNDP DRM Projects: Uttar Pradesh[23]

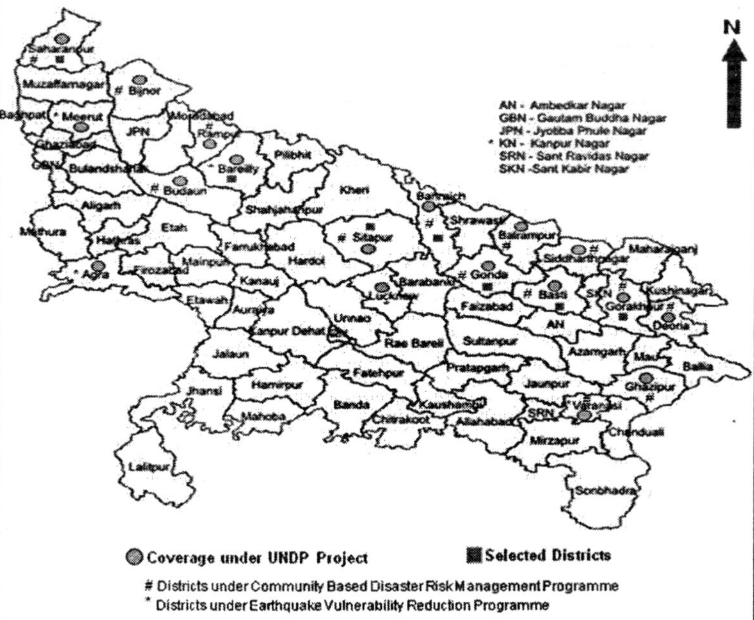

Methodology

Initially, the probable cases/success stories were identified in a sharing workshop organized in G.B. Pant Social Science Institute, Allahabad on September 8, 2008. Subsequently, a sample of 6 out of 13 districts in CBDRR project and one out of 6 districts in UEVR project were selected for the study. Information was sought on all aspects of the projects' activities in the form of write-ups, photographs, video recordings, questionnaire and reports in the print media. Each event's site was visited to conduct interviews and focus group discussions. After presentation of the draft report in a post-study workshop with all stakeholders on November 29, 2008, the final report in published form was subsequently submitted to the sponsoring organizations in January 2009. The report as per the study objectives is limited to presenting case stories rather than analyzing other aspects of the programme.

Summary Findings

The research work in all documented 31 case studies on risk management of floods, earthquakes and other disasters. These studies covered practically all components of the DRMP: development of multi-hazard risk management and sustainable recovery plans; capacity building and functioning of Disaster Management Teams (DMTs) and Committees at all levels; processes of awareness generation; special training for women in various aspects of DRM; creating and using local resource based technologies including earthquake resistant features for houses; training in retrofitting, and the construction of technology demonstration units. Out of 31 case studies covered in the larger report, here only two cases covering both a failure and a success story are presented.

Case Study-I. A New Fire Resistant Model House Not Accepted by Villagers

Barethi village in Mishrikh Tehsil, Sitapur district is inhabited by approximately 400 families. A majority of the villagers live in *kutcha* houses (mud walls and straw and bamboo roofs). On May 13, 2005, a fire flares up inside a hut and soon the entire area is ablazed. On a villager's request fire brigade arrives 90 minutes later. After grappling for twenty four hours, the fire finally comes under control, though not completely.

In this incident, one ten year old girl died, nearly fifty animals charred to death, while many others were injured. A large number of people sustained severe burn injuries and 76 houses were wrecked.

Huts Vulnerable to Fire Disasters.

The DDMT after the incident contacted the Central Building Research Institute (CBRI), Roorkee, and sought its suggestions to build fire proof *kuchcha* houses. A Civil Engineer with CBRI suggested the villagers to undergo training for building such huts. He trained the villagers for building a fireproof house involving materials like sugarcane leaves, replacing ropes, dry grass covering for the bamboo frame, a layer of coal tar mixture adding kerosene oil, cow dung and straw for wall covering. He gave a practical demonstration of the fire resistant model house.

The villagers appeared very appreciative of the new model house during the training session but none adopted it later in practice. They found the project unaffordable and beyond their capacity to bring required materials.

The Barethi case clearly depicts that:

- Externally imposed technology, deprived of local resource base is not acceptable; Only cost-effective, affordable and convenient technology is adopted by the people;
- Lessons taught in a classroom situations may falsely appear to be internalized by trainees which may become impractical in real life situations;
- All necessary variables need to be taken into account while analyzing a social problem; and
- Methods used for imparting skills for practical use must be based on real life situations and socio-economic surroundings.

Case Study-II. Women Display Courage and Interest for Developing Their Skills

June 14, 2006; a search, rescue and rehabilitation training workshop was organized in the Brahmapur block, Gorakhpur district. In all, 30 trainees from 17 villages of the block participated in the workshop, including a few women. On July 26, a mock drill was organized in a remote village, Barkachha, situated on the bank of the Rapti. Trainees were being taught how to save lives of people who are drowning. Suresh, aged 12 was asked to jump into the river. He was then saved by the participants wearing life jackets.

The women trainees wanted to practically participate in the mock drill exercise but were not allowed to do so. The women felt that they are being considered weak and vulnerable. Suddenly, Surti Devi, 45 years jumped into the river without warning and pretends to be drowned. Immediately, Singari Devi, Kunti Devi and Rina Devi, follow suit and pull her out. They also display their knowledge of administering first aid to Surti.

The men attending the training were amazed at their counterparts particularly because women had jumped into the water wearing cumbersome *sarees* and without life jackets, particularly when the river is in spate and its bank very slippery. The women showed a high level of self-confidence and proved that they are equally skilful. After their courageous feat, all the women were included in the core group of the Search and Rescue Team of the Village Disaster Management Team.

In brief, the similar stories included in the larger study reveal the following lessons:

Operating Multiple-Layers of Management: Multiple-layers of management and committees at different levels mitigated the effects of disasters. Local communities were inspired to form trained DMTs in their own regions and prepare themselves for an incumbent disaster. Trained volunteers organized themselves into various wings like vigilance team, first-aid teams, communications teams, and search and rescue teams.

Skills Acquired Practiced: Skills acquired by the community have been practiced successfully and there is demand to build further capacities. Members of DMTs displayed alertness, vigilance, swift action, sensitive preparedness that reduced risk of human lives and loss of property.

Involving the Community Paid Rich Dividends: Mobilization of resources for relief operations from the community and sensitivity of common people towards the victims paid dividends in successfully managing relief operations in some areas. Evidently, the message is that improving capability of local people is the best strategy of their empowerment.

Marginal Communities and Women Involved: Marginalized and forgotten communities were brought into the mainstream and received benefits of rehabilitation strategies and interventions. Women forcefully demonstrated a high level of self-confidence and proved that they are second to none.

Disaster Management Plans: The DRM plans from the village to district and state level were prepared and implemented to some extent. Preparation of plans was itself an educative exercise.

Network Creation: DMTs linked various associations including NGOs, trade unions, youth welfare clubs, *Mahila Samakhya*, industrial welfare associations, NCC and NSS volunteers, which made effective contributions. The students NCC/ NSS wings were provided training for skill formation through mock drills in order to prepare them to combat any future disaster.

Equipment Developed Using Local Resources: The DMTs also developed a number of inexpensive equipments using local resources for combating disasters:

- Life jackets made of cold drinks bottles and simple cloth

costing only between sixty to seventy rupees;

- bottle gourds dried and used as life jackets;
- banana trees trunks, cans of oil, *Tasla* (utensils used for cooking rice) and *Handa* (water storing utensils) used for helping people to keep afloat or used as boats;
- rolled cardboards or tin-foils used as loudspeakers for warning;
- improvised wooden ladders into makeshift commodes used for taking temporary shelter during floods;
- smokeless ovens encouraged for preventing fires;
- homemade remedies for treating patients during disasters;
- earthquake resistant houses constructed in some districts.

Minimized Crop Production Losses: In order to minimize loss in crop production in flood-affected regions, a Crop Diversity Programme had been introduced which aims at establishing a balance in crop rotation.

Success in Urban Areas: Awareness campaigns in urban areas successfully motivated people, especially in services sector. Owners and workers of tent houses made several doorways in tents where large gatherings are held; and owners of cinema halls, hotels and restaurants began using emergency gates.

Replication of DRM in Other Countries: Success of Community Based DRMP has led to its being replicated in another disaster prone country, Kenya. A visiting team from Kenya was highly impressed by the initiatives and strategies implemented in the DRMP districts.

Some More Steps Towards Success: Local people partly accepted new innovations as suggested by the Project even after massive incidence of disasters. This suggests that the experiments were either beyond their understanding or were dependent on the import of resources from outside. The successes were known at higher level as the programme was being administered and monitored at the district or city level. Information about various components of DRM was not known to many villagers.

Implementation of the post-disaster long-term rehabilitation activities appeared a little weak. In some districts land and houses of many people were destroyed due to change in river courses. Neither the district administration nor the DMTs had any concrete strategies for compensating these losses. The programme also lacked adequate provisions for documentation since its inception and there existed no system in place to identify its outcomes for improvement. In some cases, training on DRM was imparted to the stakeholders in a formal classroom like situation rather than activity based learning environment.

It was clearly noted that many districts of the state face floods and other types of disasters every year. No independent department has been formed till date for better management of these disasters. The state too has no independent monitoring evaluation and research institution for providing technical inputs for predicting disasters, for development of area specific disaster training modules and resources, and for conducting research on various socio-economic aspects of the disasters, which may provide valuable strength to DRMP.

Conclusion

Overall, the DRMP has been successful in developing a community based organizational structure for tackling disasters in Uttar Pradesh. Now mobilization of skilled human resources and volunteers has become easier at all levels. An integrated vision is emerging for making the DRMP a success with the assistance of a strong network of a variety of partners and stakeholders. In view of the rapidly changing climatic conditions, the issue of disaster risk reduction needs immediate attention of the policy-makers. Under the guidance of the Hyogo Framework of Action as mentioned earlier, a strategy based on the local conditions may urgently be evolved to enhance adaptive capacity of the communities, maximize resilience to future threats, and to minimize the growing levels of disaster risks. IPCC Report has rightly noted that the societies around the world have a long record of adapting and reducing their vulnerability to the impacts of weather and climate related events. However, in view of the rapid climate changes presently, additional adaptation measures are required at regional/local levels to mitigate its adverse impacts.[24]

The case studies clearly point out innovative ideas and lessons for mitigation of disasters which may be carried forward if shared widely with concerned stakeholders. Introduction of any preventive and innovative technology for managing disasters requires its acceptability to the local people keeping in view their socio-economic conditions and the available local resource base. An efficient communication network is of immense help during emergency in all major disaster prone zones. Well-designed post-disaster long-term rehabilitation policies appear the need of the hour for compensating disaster victims for loss of their land and shelter. Empowerment of *Gram Panchayats* and creation of a network of motivated and trained partners in each district and city would go a long way for managing emergency situations in future.

Acknowledgement

This chapter is based on the study 'Disaster Risk Management Programme in Uttar Pradesh: Learning from Case Studies', jointly sponsored by UNDP and MoHA, Government of India and Government of U.P., January, 2009. The study has been conducted and subsequently published as a Report before its submission to the sponsors by G.B. Pant Social Science Institute, Allahabad. After its presentation in IIPA, the paper has also been presented in Indian Economic Association Conference, Dec. 27-30, 2010 and published in *The Indian Economic Journal,* Special Issue, December 2010. Author expresses thanks to the sponsoring organizations for assigning the study to the Institute. Views expressed in this chapter are that of the author.

NOTES

1. ISDR (2008): Climate Change and Disaster Risk Reduction, International Strategy for Disaster Reduction (ISDR), Briefing Note-01, Geneva, September, pp. 1-2.
2. IPCC (2007). Climate Change-2007, Synthesis, Fourth Assessment Report, Valencia, Spain, November 12-17, pp. 30-33.
3. *Ibid.,* p. 30.
4. UNFCCC (1992) quoted in IPCC (2007), *op. cit.,* p. 30.
5. ISDR (2008). *Op. cit.,* p. 5.
6. UNDP (2004). Reducing Disaster Risk: A Challenge for Development, A Global Report, Bureau of Crisis Prevention and Recovery.
7. Planning Commission (2002). Tenth Five Year Plan (2002-2007), Dimensions and Strategies, Vol. I, Government of India, New Delhi, pp. 189-190.

8. ISDR (2008). *Op. cit.*, p. 5.
9. MoHA (2007). 2nd Asian Ministerial Conference on Disaster Risk Reduction: Background Note, Government of India, New Delhi, November 7-8, p. 1.
10. MoHA (2004). Disaster Management in India: A Status Report, National Disaster Management Division, Government of India, New Delhi, p. 3.
11. *Ibid.*, p. 2.
12. *Ibid.*, pp. 4-5.
13. MoHA (2002). Local Level Risk Management Indian Experience: An Initiative under the GoI-UNDP Disaster Risk Management Programme, Series-I, National Disaster Management Division, Government of India, New Delhi, p. 1.
14. Planning Commission (2008). Eleventh Five Year Plan (2007-2012) Inclusive Growth, Vol. I, Government of India, Oxford University Press, New Delhi, pp. 208-210.
15. Relief and Revenue Department (2008). Disaster and Emergency Management in Uttar Pradesh: A Status Report, Government of U.P., Lucknow, p. 1.
16. *Ibid.*, pp. 1-2.
17. *Ibid.*, p. 4.
18. *Ibid.*, p. 3.
19. *Ibid.*
20. MoHA-UNDP (2002). Disaster Risk Management Programme (2002-07) Community Based Disaster Reduction and Recovery Through Participation of Communities and Local Self Governments, Project Document, Series-2.0, Government of India, New Delhi, pp. 1-3.
21. *Ibid.*, pp. 2-3.
22. Bhatt, K.N. (2009). "Disaster Risk Management Programme in Uttar Pradesh: Learning from Some Case Studies", jointly sponsored by UNDP and MoHA, Government of India and Government of U.P., January, 2009.
23. Location Map provided by DRMP, State Project Office, Lucknow (Changes incorporated as required for showing Sample Districts).
24. IPCC (2007). *Op. cit.*, p. 56.

8

The Environment Impact Assessment (EIA) Report of the 2000 MW Lower Subansiri Hydroelectric Project

A Critical Analysis

Ranjit Dutta and *Sarada Kanta Sarma*

Introduction

Dams built on rivers pose a serious threat to biodiversity, especially when they are built on large tributaries (Roberts, 1993; Kottelat and Whitten, 1996). Apart from blocking migratory routes, dams generally alter hydrological patterns, silt deposition patterns, water temperature and water quality, leading to massive impacts on aquatic and riparian lives (Kottelat and Whitten, 1996). Natural flood regimes are an important part of aquatic ecosystems, and whenever they are altered, it can be expected to serious impacts on the downstream flood plains including stream, riparian wetlands, riparian zones, croplands and all the life associated with it (Roberts, 1993; Kottelat and Whitten, 1996).

An exhaustive Environment Impact Assessment (EIA) is utmost necessary in the downstream of the dam site of any proposed hydel project. Unfortunately, those who have planned and promoted large hydel project have rarely ever conducted in-depth investigations regarding the impacts on aquatic life, riparian life and the aquatic systems that local people depend upon for their livelihoods (Roberts, 1993; Kottelat and Whitten, 1996; World Bank, 1998).

The EIA report of the 2000 MW Lower Subansiri Hydel Project in North-east India through which the National Hydroelectric

Power Corporation (NHPC) authority got the clearance from the Ministry of Environment and Forest for construction of this large dam was very poor in respect to the downstream (Vagholikar, 2007). The reports itself claimed to have their survey done only up to a distance of 07 km (Koyala Nallah) down from the dam site and there is no information beyond this distance. However, the total downstream flood plain of the river Subansiri from the dam site (Fig. 8.1) to its confluence with the mighty river Brahmaputra is approximately 130 km, which is very rich in native aquatic macrophytes, riparian terrestrial species and indigenous crops. Subansiri river is the lifeline of the downstream riverbank people of the region because they are entirely dependent upon her for their livelihoods and day-to-day necessities. These aspects were totally ignored in the said EIA studies.

Fig. 8.1: Dam site before Construction of 2000 MW Lower Subansiri Hydel Project (2003).

In this chapter, an attempt has been made to discuss all these serious issues ignored by the NHPC authority in their EIA. A comparison in between the studies of downstream ecology of the river Subansiri and EIA report of the 2000 MW Lower Subansiri Hydroelectric Project prepared by the WAPCOS has been done in the chapter.

Methodology

The present study was conducted in the entire downstream

(approx.130 km) from the dam site to the confluence of the river (Jamuguri) with river Brahmaputra. To know the downstream river health, water quality and quantity, habitat diversity, ecological conditions of the downstream and the influences of unregulated river on the downstream agriculture productivity and livelihoods prior to the regulation of the river, were studied during 2008-2009.

For a systematic monitoring and study of each selected parameters of the downstream, it has been divided into four sectors namely Chowaldhowa, Khabolo, Dhunaguri and Jamuguri (Fig. 8.2).

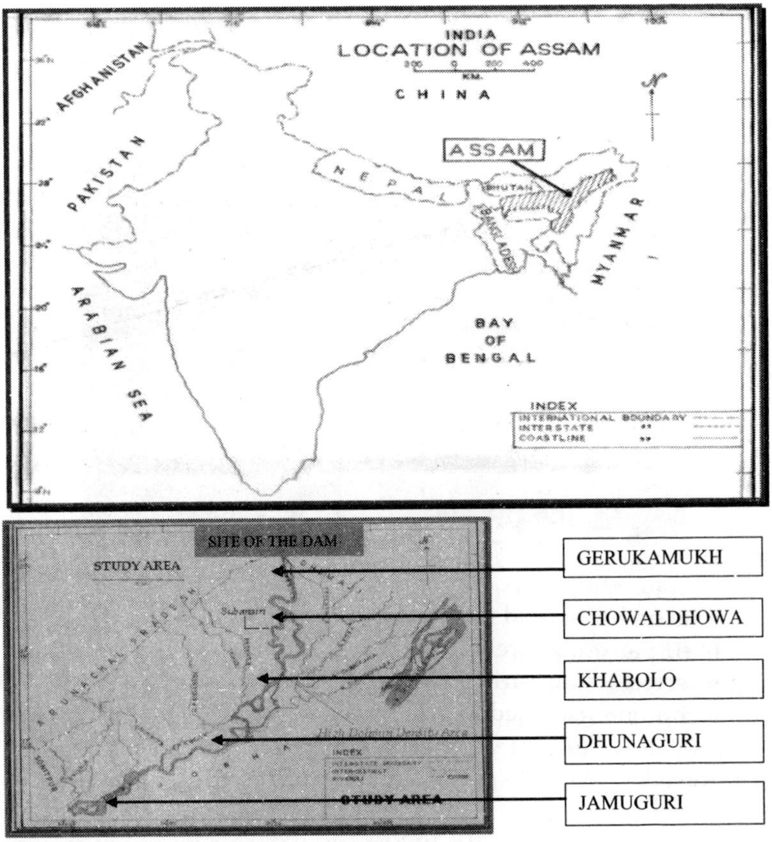

Fig. 8.2: Location Maps of the Study Area showing the Downstream of the Subansiri river.

Hydro-biological Parameters: Air and water temperature was recorded by using mercury thermometer, while transparency and pH were measured by Secchi disc and pH meter respectively. Current flow was recorded according to the method of Trivedy *et al.* (1987). Total dissolved (TDS) and suspended solids (TSS) and Dissolved oxygen were measured as per the methods of APHA (1989). Free CO_2 were studied according to the methods of Welch (1952).

Planktons Study: The qualitative analysis of the plankton was carried out according to the methods of Needham and Needham (1966), Edmonson (1974) and Lackey (1983).

Riparian Flora: Identification of macrophytes and other plant species was done by preparing herbarium sheets (Jain and Rao, 1976) besides the literature survey (Hooker, 1897; Kanjilal *et al.*, 1934-40; Bor, 1940; Dutta, 1985; Khan and Halim, 1987; Cook, 1996).

Fish Diversity: The specimens were identified according to the methods of Munshi & Srivastava (1988), Talwar & Jhingran (1991), Biswas (1993) and Viswanath (2002).

River Dolphin Survey: By surveying the entire downstream with motorboat, the dolphins were counted according to the methods as described by Mohan *et al.* (1998).

Avian Fauna: Residential and migratory birds were recorded in the different habitats of the downstream flood plain by surveying, photography and literature survey (Chowdhuri, 2000).

Agricultural Diversity: The crop diversity was recorded by surveying the downstream flood plain. The indigenous deep-water rice varieties were collected and preserved.

People's Livelihood: Information recorded in this study was through the direct observation and discussion with riverine communities, besides using a prepared questionnaire to solicit for data on household socio-economic characteristics and floodplain used patterns in relation to the influence of river ecology in the selected areas during 2008-2009. A stratified random sampling technique was used to select 120 households for the interviews.

Results

The findings of the EIA report and the present investigation have been summarized below:

(A) Downstream Information from the EIA Report (WAPCOS)

Total Study Area Reported	:	07 km on either side of the river up to the intersection of Koyala Nallah.
Water Quality Parameters	:	Appropriate locations within the study area.
Land use Pattern	:	No record in the downstream.
Terrestrial Ecology	:	Only up to Gerukamukh.
Total Numbers of Species	:	Tree: 26, Bamboo: 02, Shrub: 14, Herb: 24.
Species Diversity Index	:	Tree: 1.37, Bamboo: 0.32, Shrub: 1.03, Herb: 1.29.
Evenness Index	:	Tree: 0.97, Bamboo: 1.00, Shrub: 0.90, Herb: 0.93
Faunal Species	:	List of faunal species reported in the area: Mammal: 10, Snake: 10, Birds: 13, Insect: 15.
Aquatic Ecology	:	Only up to Gerukamukh.
Total no. of Species	:	Phytoplanktons: 08.
Fish Diversity	:	Below 800 m MSL.
Total number of species	:	27.

Prediction of Impact: (During the construction phase)

a. Soil Erosion: According to the EIA report, soil erosion in different construction phases will create adverse impacts on photosynthetic action and biological productivity. However, no mention has been made about the impacts on other aspects of downstream ecology.

Management Measures: *Not mentioned*

b. Muck Disposal: A large quantity of muck will be generated because of tunneling operations, construction of roads, etc. The total quantity of muck to be generated is of the order of 81.5 lakh m^3.

Management Measures: A total area of 78 ha has been earmarked for muck disposal, of which 40 ha lies in Arunachal Pradesh. The remaining, 38 ha lie in Assam. In the proposed project, two of the three muck disposal sites are along rivers as well. Hence, in the proposed project, adequate measures such as retaining walls, etc. need to be constructed to ameliorate the likely adverse impacts (Fig. 8.3).

Fig. 8.3: Damping of Muck into the Downstream of the Subansiri River

Impacts on Water Quality

No mention on the changes in flow regimes and its impacts on other aspects of downstream ecology have been made.

Impacts on Terrestrial Flora

No information on the riparian flora of the downstream.

Impacts on Terrestrial Fauna

No information of terrestrial fauna of the downstream.
No information about the elephant corridor in the downstream.

Impacts on Avian Fauna

No information about residential and migratory birds in the downstream.

Impacts on Aquatic Ecology

(a) Construction Phase

Report says that the extraction of gravel and sand causes considerable damage to fish stocks and other aquatic life by destabilizing the sub-stratum, increasing the turbidity of water, silting of the channel bottom and modifying the flow, which in turn may result in erosion of the river channel. These alterations upset the composition and balance of aquatic organisms. During construction of a river valley project, huge quantity of muck is generated at various construction sites, if not properly disposed, would invariably flow down the river during heavy precipitation. This can have adverse impacts on the development of aquatic life. The inadequate system for dumping of debris can lead to many undesirable impacts. Unfortunately, no mention about the impacts on other aspects of the downstream aquatic ecology has been made in the EIA report.

Management Measures: *Not mentioned.*

(b) Operation Phase

Alteration of the habitat would bring changes in physical and chemical state and the biotic life. It can be expected that the most important changes are the reduced flow rate, increase in water temperature, and reduction in availability of stenothermal aquatic animals, increase in population of fury-thermal species. No mention of the impacts on other aspects of the downstream aquatic ecology has been made.

Management Measures: *Not mentioned.*

(B) Observation of the Present Investigations

The observations are summarized as follows:

1. *Water Quality:* The pre-dam water quality of the downstream river is suitable for the aquatic biota (Table 8.1).

The downstream up to a length of 130 km provides other different habitats like floodplain wetlands, meandering cut-off etc.

2. A good number of phytoplankton and zooplanktons species have been recorded from the downstream (Table 8.2).

3. Downstream provides unique habitat for fresh water fishes, out of which some are endangered, rare or threatened. Ornamental fishes are also in abundance. So far, 168 fish species were collected

Table 8.1: Water Quality in the Downstream of Subansiri River during 2008-2009

Area	*Sector I*			*Sector II*			*Sector III*			*Sector IV*		
	M	*SE*	*SD*	*M*	*SE*	*SD*	*M*	*SE*	*SD*	*M*	*SE*	*SD*
Air temperature (°C)	25.8	1.73	6.02	24.8	1.47	5.09	24.17	1.51	5.23	24.1	1.55	5.39
Water Temperature (°C)	18.52	1.23	4.28	18.7	1.23	4.26	18.5	1.41	4.9	18.55	1.26	4.39
Current flow (m/s)	2.6	1.1	1.92	2.3	1	1.73	2.5	1.2	1.79	2.4	1	1.71
	(H)	(L)	(AV)	(H)	(L)	AV	H	L	AV	H	L	AV
Transparency (cm)		5.47	18.97	42.29	5.69	19.72	43.9	5.52	19.12	45.32	6.08	21.08
pH	7.44	0.09	0.33	7.55	0.1	0.38	7.53	0.1	0.35	7.58	0.1	0.37
Dissolved oxygen (mg/l)	8.61	0.23	0.82	9.7	0.3	1.05	10.33	0.32	1.14	9.86	0.47	1.64
Free carbon dioxide (mg/l)	4.71	0.3	1.04	4.98	0.39	1.31	5.2	0.35	1.22	4.25	0.22	0.78
Total dissolved solids (ppm)	162.41	3.26	11.31	166.25	4.22	14.64	165.08	4.61	16	166.83	4.6	15.93
Total suspended solids (ppm)	29	1.84	6.38	28.91	2.28	7.91	29.75	2.22	7.7	29.41	1.04	3.62
Total solids (ppm)	191.41	4.85	16.83	195.16	5.8	20.11	194.83	5.51	19.11	196.25	5.23	18.15

At the 0.05 level, the means are significantly different.

Key: M = Mean; SE = Standard Error and SD = Standard Deviation, H = Highest, L = Lowest, AV = Average
Sector I: Chowaldhowa, Sector II: Khabolo, Sector III: Dhunaguri, Sector IV, Jamuguri.

from the study areas and preserved which includes the downstream river course and the flood plain wetlands (Table 8.2).

4. More importantly, the downstream has a good population of endangered aquatic mammal, the Ganges river dolphin (Table 8.2).

5. Flood plain wetlands in the downstream are rich in aquatic plants from where 123 aquatic macrophytes have been recorded (Table 8.2).

Table 8.2: Biodiversity Status of the Subansiri River during 2008-2009

1. Downstream from the dam site	130 km (Approx.)
2. Flood plain/Riparian Wetlands	98 numbers
3. Phytoplankton (No. of species)	48
4. Zooplankton (No. of species)	23
5. Avian Fauna (No. of species)	25
6. Fishes (No. of species)	168
7. Numbers of River Dolphin	26
8. Macrophytes (No. of species)	123
9. Grasses (No. of species)	119
10. Indigenous deep-water rice varieties.	23

6. Twelve giant macrophytes (height is more than 1.5 meter) are recorded, those are amphiphytic in nature and unique in the downstream riverbank having tremendous economic and ecological values.

7. River riparian zones are composed of large number of trees, shrubs, climbers, herbs and grass.

8. Downstream people's livelihoods are related with the wild nature of the river. All together seven different livelihood sectors have been identified on which people of the study areas are dependent and the free flowing Subansiri River influences all (Table 8.3).

9. In the downstream flood plain, 23 indigenous deep-water rice varieties have been recorded, identified and preserved.

10. Downstream of the river and the floodplains are the major habitats of large numbers of migratory birds.

Conclusion

The EIA report, through which the NHPC got the clearance from Ministry of Environment and Forest, Govt. of India, possesses many loopholes. It becomes evident that no sufficient and detailed

Table 8.3: Influences of Subansiri River on Different Livelihood Sectors

Sl. No.	*Influence of river on different livelihood sectors of downstream people*	*Benefits*
1.	Agriculture in the downstream of Subansiri River	Food/Income
2.	Collection of wood from the downstream of Subansiri River	Income
3.	Livestock (Buffalo & Cattle) rearing of the downstream of Subansiri River	Income
4.	Fisheries of downstream of Subansiri River	Food/Income
5.	Transportation and trade in the downstream of Subansiri River	Income
6.	Riparian economic flora of the downstream of Subansiri River	Economic values/ income
7.	Sand and Gravel mining in the downstream of Subansiri River	Income

study was done in the entire downstream during the preparation of the report. Detailed report mentions about the probable impacts upon the downstream river health, rare and endangered species, physical resource, people's livelihood and the options for mitigation, etc. did not find place in the report. The EIA report contains insufficient records on terrestrial and aquatic flora, phytoplankton, fish diversity, riparian flora, zooplanktons, avian fauna, floodplain crops, people's livelihood, etc. besides healthy habitats controlled by the unregulated river Subansiri. Interestingly no report on the IUCN red listed and national aquatic animal of India, Ganges dolphin *Platanista gangetica* Roxb. has been made in the report, but still the Subansiri river provides a good habitat for this animal. These ecologically rich downstream with existing environmental scenario might be adversely affected due to the construction and operation of the project as referred earlier and chances of elimination of endangered, threatened and native species through habitat fragmentation. The most significant adverse effect will be on the livelihood of downstream people. River Subansiri and the flood plain wetland ecosystem provide many services that contribute to peoples' well-being and poverty alleviation. These include provision of food and fiber, water, building materials and medicinal plants that help to meet the basic human needs, provision of earning cash through transportation business, live stalk rearing, drift wood collection, agricultural product and most essential is the fishing. The degradation of river and other wetland ecosystems due to dam construction can have profound economic and social implications.

From the above discussion, it is suggested that:

A. Consider the issues of the total downstream river health assessment.
B. Establish the likely sedimentation effect on the downstream. Assess the fishery value of the downstream to the dam site of the river Subansiri.
C. Establish the status, ecology and range of the habitat of the endangered river dolphin in the downstream and likely impact of the dam on the habitat and distribution of the river dolphin.
D. Assess the probable impact on groundwater table on the downstream flood plain due to reduction of water discharge for damming off the river.
E. Assess the effect on the livelihood activities of the downstream people.
F. In-depth studies with documentations are necessary before taking decision on operation of the 2000 MW Lower Subansiri hydroelectric project.

Acknowledgement

The authors are thankful to Dr Debojit Baruah of Lakhimpur Girls' College, Assam for help in phytoplankton and plants collection and identification, Dr Shyma Prasad Biswas of Dibrugarh University for help in fish and zooplankton identification, Sri Lakhi Prasad Hazarika of North Lakhimpur College, Assam for help in river dolphin survey and bird's identification. The first author is also thankful to University Grants Commission for providing financial support as a UGC Project Fellow at Lakhimpur Girls' College, Assam.

REFERENCES

APHA, (1989). *Standard method for the experimentation of water and wastewater,* American Public Health Association, New York.

Biswas S.P. (1993). *Manual of Methods in Fish Biology,* A South Asian Publishers Pvt. Ltd., New Delhi.

Bor N.L. (1940). *Flora of Assam,* Govt. of Assam, India, Vol. IV.

Chowdhuri A. (2000). *The Birds of Assam,* Gibbons Books, WWF-India.

Cook C.D.K. (1996). *Aquatic and Wetlands Plants of India,* Oxford University Press, India.

Dutta A.C. (1985). *Dictionary of Economic and Medicinal Plants,* Assam Printing Works: Jorhat, Assam, India.

Edmonson W.T. (1974). *A manual on methods for measuring production in aquatic Environment.* Academic Press, New York.

Hooker J.D. (1897). *Flora of British India,* Reevs & Co., London. Vol. I-VII.

Jain S.K. and Rao R.R. (1976). *A handbook of field and Herbarium methods,* Today and Tomorrow Printers & Publishers: New Delhi.

Kanjilal U.N. Kanjilal P.C. and Das A. (1934-40). *Flora of Assam,* Govt. of Assam, India, Vol.: I-IV.

Khan M.S. and Halim M. (1987): *Aquatic Angiosperms of Bangladesh,* Bangladesh National Herbarium.

Kottelat M. and Whitten A.J. (1996): "Freshwater biodiversity in Asia with special reference to fish." *World Bank Technical Paper* No. 343, Washington DC., i-ix, 1-59.

Lackey J.B. (1983): The manipulation and counting of river plankton and changes in some organism due to formalin preservation. Publ. Health Deptt., 53: 2080.

Mohan R.S.L., Dey S.C. and Bairagi S.P. (1998): "On a residential population of the Ganges river dolphin Platanista gangetica in the Kulshi River (Assam) a tributary of Brahmaputra." *J. of Bombay Natural Hist Soc.,* 95(1):1-6.

Munshi J.S.D and Srivastava M.P. (1988): *Natural Histories of Fishes and Systematic of Freshwater Fishes of India,* Narendra Publ. House, New Delhi.

Needham J.G. and Needham P.R. (1966): *A guide to study of freshwater biology,* Holden day Inc. San Francisco, p. 180.

Roberts T.R. (1993). "Just another dammed river? Negative impacts of Pak Mun on fishes of Pak Mun Dam on fishes of the Mekong basin." *Nat. Hist. Bull.* Siam Soc., Bangkok, 41: 105-133.

Talwar P.K. and Jhingran A.G. (1991): *Inland Fisheries of India and Adjusting Countries,* Oxford & IBH Publ. Com., New Delhi, vol. 1 and 2.

Trivedy R.K. Goel P.K. and Trisal C.L. (1987): *Practical Methods in Ecology and Environmental Sciences,* Enviro Media Publication, Karad, India. pp. 123-126.

Vagholikar N. (2007): "Lower Subansiri Hydel Project: Stop all work until critical issues are resolved." *Seuj Cinta, The News Letter of Green Heritage,* Vol.: IV, 17-22, Lakhimpur, Assam, India.

Viswanath W. (2002): *Fishes of North East India, A field guide to species Identification,* NATP, NBFGR, Lucknow.

Welch P.S. (1952): *Limnology,* McGraw Hill Book Company, Inc., New York.

World Bank, (1998): "Integrating freshwater biodiversity conservation with development: some emerging lessons." *Environment Department Papers* No. 61, Washington, DC.

9

Fresh Water Resources in India

Issues and Challenges

Amit Kumar Singh

Introduction

Water is the most precious resource that sustains all life on Earth. Leonardo da Vinci termed water as the driver of nature. Water is vital to all forms of life and fundamental for environmental health and management. It has no substitute for most uses and it will continue to be a vital source for drinking water, industrial manufacturing, power generation and agricultural irrigation. Although water covers 71 per cent of earth surface, the availability of fresh water is very limited. About 97 per cent of the earth's water is in the oceans, which is unfit for human consumption and other

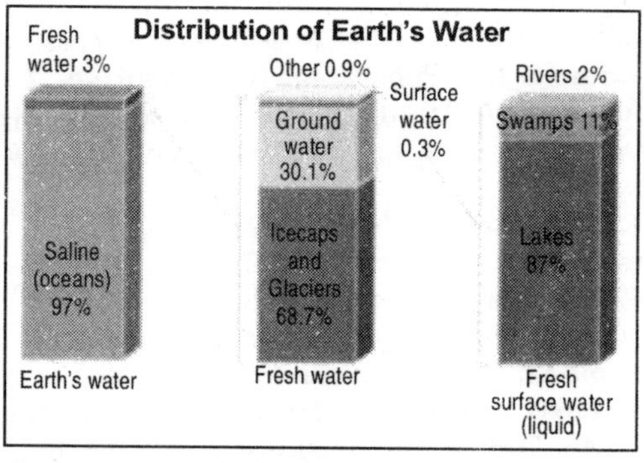

Fig. 9.1: Distribution of Earth's Water.

Source: World Water Resource Institute.

use because of its high salt content. A very small amount of salty water is also located in saline lakes (e.g., the Caspian Sea and Aral Sea). The remaining water inventory (3%) is fresh water. Out of this total fresh water resource 68.7 per cent (2% of total fresh water) is locked in the polar ice caps; and the remaining 30.7 per cent (0.9% of the total fresh water) constitutes all the fresh water in hydrological cycle including groundwater reserves. Only 0.3 per cent (0.1% of the total fresh water) is available as fresh water in rivers, lakes, and streams, which is suitable for human consumption. Water is a replenishable but a finite natural resource and it can be only renewed through the process of the water cycle. Therefore, availability of water for the society is very limited and whatever will be our demand has to be fulfilled by this limited amount. Unfortunately, human beings are not only putting tremendous pressure on the available water resources due to increasing population and growing consumption but also disrupting the natural water cycle through pollution induced climate change.

Since the beginning of the 20th century, the world's population has tripled from 2 billion to 6 billion however, the use of renewable water resources has grown six-fold (World Water Council). According to the World Water Council, 1.1 billion people are currently living without clean drinking water, while another 2.6 billion lack adequate sanitation. It is estimated that by the year of 2050, the number of people on Earth will increase from 6.5 billion to about 9 billion. Right now when 54 per cent of the annual available fresh water is already being used, therefore, to support 3 billion more people we will have to extract more groundwater resource, construct dams on river, recycle waste water and develop efficient technology. This increase in number of people on Earth would use 95 per cent of the available fresh water. This is not taking into account the needs of those people who currently do not get safe drinking water. The effects on the ecosystems would be devastating. Springs, rivers and other wetlands as well as underground water sources would run dry; lakes and estuaries where fish stocks breed would be irreversibly damaged; the list of impacts would be long.

As far as India is concerned, it is the second largest populous country of the world and contributes 16 per cent to the world total population. Despite that, the current demographical trends for

India are alarming. Every estimate for India's population growth indicates that by the year 2025 it will become 1.35 billion and by the 2050 it will become 1.65 billion. India will surpass the China population and would become the most populous country of the world. However, it is important to note that population of India is bound to increase but the amount or availability of fresh water actually available to people is finite. To feed this population we have to produce more food grains. But problem is that we have limited area of agricultural land therefore to increase our food production either we have to increase our cultivable agricultural land or to raise our agricultural productivity. In both cases we need more water. Apart from that in the coming decades, the demand for fresh water from the domestic and industrial sectors is expected to increase manifolds as more and more people are moving towards urban centres. Again the process of industrialization is also accelerating which can lead India towards an acute fresh water resources crisis. Under this background, the present chapter attempts to study the impact of growing population and changing socio-economic practices on fresh water availability in India.

Table 9.1: Water Scenario in India

S. No.	*Items*		*Quantity (cu.m.)*
1.	Annual Precipitation (Including snowfall)		4000
2.	Average Annual Availability		1869
3.	Per Capita Water Availability (2001) in cubic meter		1820
4.	Estimated Utilizable Water Resources		
	(i) Surface Water Resources	690 Cu.Km.	1123
	(ii) Ground Water Resources	433 Cu.Km.	

Source: National Institute of Hydrology, 2005.

Objective

The main objective of this chapter is to assess the impact of population growth, agricultural practices, urban water demand and waste generation, Industrial water pollution on our fresh water.

Availability and Consumption of Water Resources in India

India which has 16 per cent of the world's population and 2.45 per cent of the world's land area possesses 4 per cent of the fresh water resources of the world. With an average annual rainfall of 1,170 mm, India is one of the wettest countries in the world. India receives 4000 billion cubic metre of rainfall per year, however the availability from surface water and replenishable groundwater is only 1869 billion cubic metre (Table 9.1). Though the average rainfall is adequate, nearly seventy five per cent of the rain pours down in less than 120 days, from June to September. Apart from that due to certain limitations imposed by physiography and topography of India as well as the present state of technology to harness water resources economically restrict the amount of potential utilizable surface water of the country. According to an estimate made by the Central Water Commission, the utilisable water resources through surface structures is about 690 cubic km. and the utilisable groundwater is about 432 cubic km. Thus, total utilisable water resource is assessed as 1122 cubic km.

The above data indicates that the availability of fresh water in India is abundant, but they are unevenly distributed among and within the states. The annual rainfall varies from about 100 mm in western region to about 2,500 mm in north-eastern parts of the country. Actually, the monsoonal weather patterns of India have a major influence on the spatial and temporal variation of water availability within the country. Despite that, on an average India receives 1170 mm rainfall annually. In terms of per capita availability of annual renewable water, it was 8192 m^3 per year at the beginning of last century and at the time of Independence it was 5694 m^3. Since then, the population of India increased manifolds which decrease the per capita availability of annual renewable water continuously. Table 9.2 compares the population growth with the per capita availability of annual renewable water for India. One can see from the table that the exponential growth in population since 1950s had reduced the per capita availability of annual renewable water. It was 5177 m^3 in 1951, 1820 m^3 in 2001. It is further expected to come down to 1341 m^3 in the year 2025 and 1140 m^3 by 2050 (Table 9.2).

According to the United Nation's MDG definition an area can be defined as water stress when annual water supplies drop below

1700 m³ per person. When annual water supplies drop below 1000 m³ per person, the population faces water scarcity, and below 500 cubic metres "absolute scarcity". If we evaluate the required amount of annual renewable water for India with UN Millennium Development Goal that we can say that India is moving towards a water stress region from a water rich region.

Table 9.2: Per Capita Availability of Water in India

Year	*Population (million)*	*Per Capita Water (m^3)*
1901	238	8192
1951	361	5177
1991	846	2209
2001	1027	1820
2025*	1330	1567
2050*	1532	1360

Source: National Institute of Hydrology, 2005.
* Projected Data.

In fact the availability and distribution of water resources had not only determined the evolution and expansion of the civil society in India but would also determine the future of Indian society and economy. As per the Julius and Buky, "the development of a country's water resources is important both as an input to the economic growth process and as an element of the basic needs package which most countries strive to provide for their population". Apart from drinking, and irrigation, water is a very important factor for industrial development, power generation, transportation, aquaculture and other domestic purposes.

In a developing country like India, agriculture sector is the predominant user of water. About 700 million population are still dependent on agriculture. Therefore, the availability of water plays an important role in the rural economy as it not only provides food security to the people but also is the major source of employment. In many regions, e.g. Punjab, Haryana, Western Rajasthan and Eastern UP, irrigated agriculture has played a major role in the development of rural economies, supporting economic growth and poverty reduction. At present agriculture sectors are consuming 78 per cent of total annual water supplies of India and as per the estimates of National Institute of Hydrology, the demand for water will not reduce significantly in 2025 or 2050

(Table 2.3). At the same time, the demand for fresh water for domestic water will increase considerably as more and more people will reside in urban areas. Right now 28 per cent (Census 2001) of Indian Population is living in urban areas; this will rise to 605 million in 2025, and 971 million in 2050 respectively, indicating large demand for the fresh water (Table 9.3).

Table 9.3: Annual Requirement of Fresh Water for Different Uses (km³)

Uses	*2010*			*2025*			*2050*		
	Low	*High*	*Per cent*	*Low*	*High*	*Per cent*	*Low*	*High*	*Per cent*
Irrigation	543	557	78	561	611	72	628	807	68
Domestic	42	43	6	55	62	7	90	11	9
Industries	37	37	5	67	67	8	81	81	7
Power	18	19	3	31	33	4	63	70	6
Inland Navigation	7	7	1	10	10	1	15	15	1
Environment Ecology	5	5	1	10	10	1	20	20	2
Evaporation Losses	0	42	6	50	50	6	76	76	7
Total	652	710	100	784	843	99	973	1080	100

Source: National Institute of Hydrology, 2005.

Other sectors which would require more supply of fresh water are industrial sector, power sector, inland navigation and other ecological sectors (Table 9.3). It is quite obvious that the increase in demand for annual water supplies for these sectq s will ultimately put constraint on agricultural sectors. It is so because the total utilisable water resource in India is estimated to be only 1122 cubic km. Therefore, the capacity of agriculture sector to maintain current per capita food production will definitely be affected. And at the same time, the poor people of India with low income will have to suffer mostly from highly inflated food prices.

Stress on Groundwater Resources

Besides surface water, groundwater is the major source of drinking water in both urban and rural India. It is an important source of water for the agricultural and the industrial sector. In general, the quality of underground water is found to be higher than surface water and requires little or no treatment to make it fit for consumption. This is one of the main reasons that many of the human settlements throughout the world survives around groundwater resources. According to the Central Ground Water Board report, India's total annual replinishable underground fresh water is 433 BCM (Billion Cubic Meters)—344 BCM through natural recharge from rainfall and 89 BCM from canal irrigation system. The in-storage groundwater reserve for the country is approximately 10,812 BCM.

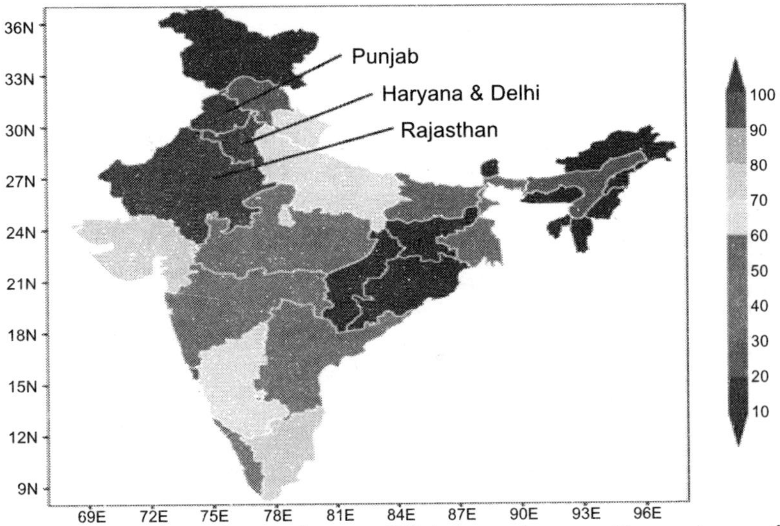

Fig. 9.1: India's Groundwater Withdrawals as a Percentage of Groundwater Recharge.

Source: NASA/Matt Rodell.

It is notable that India is utilizing 58 per cent of its groundwater resource. However, there are few states where the status of groundwater development is comparatively very high, e.g. in the states like Delhi, Haryana, Punjab, Rajasthan, UTs of Daman & Diu and Pondicherry, the state of groundwater development is more than 100 per cent (Fig. 9.2), which implies that in these states, the average annual groundwater consumption is more than average annual groundwater recharge. In the states of Gujarat, Karnataka, Tamil Nadu and Uttar Pradesh, the average state of groundwater development is 70 per cent and above. In rest states/ UTs the stage of groundwater development is below 70 per cent.

Groundwater Stress Areas /Regions of India

In 2004, Central Ground Water Board of India carried out a study in 5831 Blocks/Mandals/Talukas of all India to find out the Dynamic Ground Water Resources of India. As per the Report, 310 Blocks/ Mandals/Talukas out of a total of 5831, fall under 'overexploited' category (Table 9.4) (utilization >100% of replenishable groundwater) and 160 Blocks/Mandals/Talukas/Watersheds fall under 'dark' category (utilization >85% but < 100%) in 13 States of the country, namely, Andhra Pradesh, Bihar, Delhi, Gujarat, Haryana, Karnataka, Kerala, Madhya Pradesh, Maharashtra, Punjab, Rajasthan, Tamil Nadu and Uttar Pradesh. In number of blocks, groundwater has been overexploited and dark blocks have increased from 428 in 1995 to 470 (1998-99) and this trend continues. Approximately 12 per cent of all aquifers are severely overdrawn. So, as the groundwater levels fall, the cost of groundwater extraction is increasing as drilling becomes more expensive for the deeper extraction. According to a Government of India 1989 statistics, the cost per hectare of irrigation (at 1970-1971 prices) rose from about Rupees 3300 in the First Plan (1951) to Rupees 5400 in the Seventh Plan (1986) period.

Table 9.4: Categorization of Groundwater Stressed Blocks/Mandals/ Taluks in India

	Total No. of Assessed Units	*Safe*		*Semi-Critical*		*Critical*		*Over-exploited*	
		Nos.	*Per cent*	*Nos.*	*Per cent*	*Nos.*	*Per cent*	*Nos.*	*Per cent*
India	5723	4078	71	550	10	226	4	839	15

Source: Central Ground Water Board Report, 2006.

It is estimated that by 2050 AD, 22 per cent of the geographical area and 17 per cent of population of the country will be under absolute scarcity conditions having access to water availability of less than 500 cu m per year and 70 per cent of the area and 76 per cent of population will have access to water availability of less than 1000 cu m per year, which is identified as stress or scarcity level. However, it is notable that groundwater is the most important source of all domestic water consumption of India as around

80 per cent of drinking water demand of rural area and 50 per cent of urban areas are fulfilled by groundwater. Together with that groundwater already provide 55 per cent of irrigation water requirement and the demand for water from industry is also increasing with opening up of new enterprises driven by decade-old economic reforms. All these indicate a very dismal picture of India's water future. To overcome such problems comprehensive strategies are required to be adopted.

Pollution Threat to Fresh Water Resources

In general nature has its own capacity to renew the earth's polluted water stocks that occurs time to time naturally. Unfortunately, after industrial revolution and explosion in population growth of 20th century, many of the activities of mankind (domestic use, industry or agriculture) are now polluting watercourses and water tables dangerously. At present mankind is generating that much of polluted water which is beyond the renewing capacity of Earth's natural system. They are not only damaging the ecosystem but also putting a question mark on the survival of millions of population. There are a few natural sources of pollutants present in aquatic ecosystems. But mostly, freshwater ecosystems may become unbalanced by factors due to human activities. Human activities affect the bioavailability of chemicals to organisms, cause temperature fluctuations, and modify rainfall, pH and salinity. India is no exception to that. Most of its surface water resource is being polluted by the dumping of domestic, industry or agriculture waste water.

According to the 2001 Census, more than 22.1 per cent (230 million) of population do not have access to clean and fresh drinking water. Figure 9.3 shows the source of drinking water for rural and urban households of India. Out of these, 230 million citizens who are without clean fresh water for domestic use, 10 per cent is urban population (28 million out of total of 280 million) and 26.8 per cent is rural population (202 million out of total 740 million). Central Water Commission estimates that almost 90 per cent of the country's water resources are polluted with industrial discarding, domestic waste and pesticide and fertilizer run-off from agriculture sector. The problem of water pollution can be attributed to many factors. But the major sources of water pollution can be classified as municipal, industrial and agricultural.

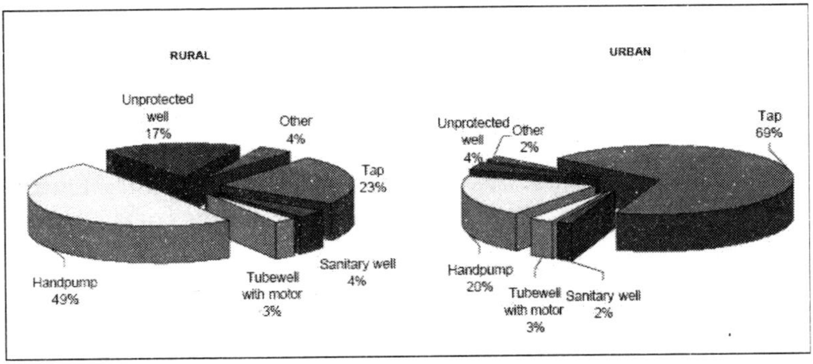

Fig. 9.3: Pattern of Water Consumption in India.

Source: Census of India, 2001.

Municipal Wastage

Municipal wastes are commonly referred to as refuse that are generated by the households of different size of urban centres. India is moving towards urban society and urban population is growing at a very high rate, and more and more rural settlement are increasingly being urbanized. There is sufficient empiric and statistical evidence that water usage per capita is higher in the urban areas. But an alarming fact is that the rate of waste water generation increases with increased disposable household incomes. Together with that a large number of the cities/towns either do not have any sewerage system or the sewerage system is overloaded or defunct. All this results in large amount of wastewater uncollected. The bulk of pollution shall get retained on land to percolate, leach or get washed-off to streams or groundwater. By looking at the present capacity and performance of the present waste water treatment plant, it is doubtful whether India would be capable of recycling huge domestic sewage from urban settlement. At present, domestic sewage from cities and towns is the biggest source of pollution of water bodies in India. According to the OSEC, Business Network Switzerland in India, the infrastructure for the treatment of used or wasted water is inadequate, only about 26 per cent of domestic and 60 per cent of industrial wastewater is treated. In 423 class I cities (i.e. cities with a population of more than 1,00,000), only 29 per cent of wastewater is treated (Table 9.5). Class II towns (i.e. towns with a population between 50,000 and 1,00,000) are able to treat just 4 per cent of wastewater. Even India's national capital, New Delhi, treats less than half of the 3,267 million litres of wastewater it generates every day.

Table 9.5: Status of Wastewater Generation, Collection and Treatment in Class I Cities and Class II Towns (MLD)

Type	*Number of Cities/ Towns*	*Waste water Generated*	*Water Collected*		*Water Treated*		
			MLD	*Per cent (of generated)*	*MLD*	*Per cent (of collected)*	*Per cent (of total)*
Class I cities	299	16662.5	11938.2	72	4037.2	33.8	24
Class II towns	345	1649.6	1090.3	66	61.5	5.6	3.7
Total	644	18312.1	13028.5	71	4098.7	31.5	22.4

Source: CPCB 2000.

According to the Census of India, the population of India is likely to be stabilized by 2050 at the level of 1700 million people. At present, the urban population is 285 million and keeping in view the population projection for the year 2051 it is likely to be of the magnitude of 1093 million. The per capita wastewater generation is around 121 litre/capita/day based on the average wastewater generation observed during the three studies carried out by CPCB. Based on the projected population for the year 2051, the wastewater generation is going to be around 1,32,000 mld. This large amount of partly treated or untreated wastewater from urban centre will be disposed into natural water resources. In view of such situation, there is a need to attain 100 per cent wastewater treatment in each city with more stringent standards.

Industrial Effluents

Economic reforms of 1990s contributed to a steady and impressive growth in Industrial sectors in India, which has contributed to high economic growth but simultaneously it has also given rise to severe environmental pollution. Although, industrial sector accounts for only 3 per cent of annual water withdrawals in India, its contribution through the effluent load is disproportionate to its consumption figures. These effluents are often contaminated with highly toxic organic and inorganic substances which remain in the ecological systems for many years. Most of these industrial effluents are discharged into rivers without being treated. Industrial waste water usually contains specific and readily identifiable chemical compounds.

Table 9.6: Projected Population and Respectively Wastewater Generation

Year	*Urban Population*	*Litres/Capita/ Day (lpcd)*	*Gross Wastewater Generation (mld)*
77-78	60	116	7007
89-90	102	119	12145
94-95	128	130	16662
2001	285	-	-
2011	373	-	-
2021	488	121 (Assumed)	59048 (Projected)
2031	638	121 (Assumed)	77198 (Projected)
2041	835	121 (Assumed)	101035 (Projected)
2051	1093	121 (Assumed)	132253 (Projected)

Source: CPCB 2000.

During the last fifty years, the number of industries in India have grown rapidly. But water pollution is concentrated within a few subsectors, mainly in the form of toxic wastes and organic pollutants. Out of this, a large portion can be traced to the processing of industrial chemicals and to the food products industry. In fact, a number of large and medium-sized industries in the region covered by the Ganga Action Plan do not have adequate effluent treatment facilities. Most of these defaulting industries are sugar mills, distilleries, leather processing industries, and thermal power stations. Most of the major industries have treatment facilities for industrial effluents. But this is not the case with small-scale industries, which cannot afford enormous investments in pollution control equipment as their profit margin is very slender.

According to the CPCB in 2000, it is estimated that 13,500 million litres per day (MLD) of Industrial wastewater is generated from different industries. However, the treatment capacity available for industrial wastewater was only for 8,000 MLD. Thus, there is a big gap (5,500 MLD) in treatment of industrial wastewater, mainly generated from small-scale industries. It is notable that most of these small scale industries are located in residential areas. In these areas, establishing effluent treatment systems is a big problem because they have space constraint. Moreover, the small-scale industries are not having adequate resources to establish treatment systems.

Agricultural Practices and Water Problem

Agriculture is the single largest user of freshwater on a global basis and India is not an exception. As mentioned earlier agriculture is utilizing 78 per cent of the fresh water resources of India at the same time this sector is the major cause of degradation of surface and groundwater resources. Before 20th century when agriculture was based on natural manure and local irrigation system Earth had a capacity to recycle back to surface water and/or groundwater. Now agriculture practices have changed completely. It is more mechanized and is very much depended on hybrid seeds that need lots of water from irrigation, fertilizers and pesticides. Therefore, now agriculture is both cause and victim of water pollution. It is a cause through its discharge of pollutants and sediment to surface and/or groundwater, through net loss of soil by poor agricultural practices, and through salinization and waterlogging of irrigated land. It is a victim through use of wastewater and polluted surface and groundwater which contaminate crops and transmit disease to consumers and farm workers (FAO).

According to United Nations, by the year of 2025, world will require 40-45 per cent more food production due to large increase in global population. Therefore, the role of irrigation agriculture will be increased very much to raise the global food supply. But its **side effect** should also be kept in mind. At present agriculture sector is one of the major factors of water pollution through salinization process, use of fertilizers and pesticides. In case of India since 1951, the percentage of irrigated land, use of fertilizers and pesticides increased manifold and so the level of water pollution. According to a study done by J. Rank in India more than 4 million hectares of high-quality land have been abandoned because of salinization and waterlogging caused by too much irrigation.

The use of fertilizers in India has risen from 7.7 million tonnes to 13.9 million tonnes in 1993-94, an further increase to 19.2 million tonnes in 2001. Though this increase in fertilizer helped India to increase the food production but the fact of matter is that fertilizers have also ended an indirect adverse impact on the water resources. Importantly as they have high nutritional content they allow organisms to proliferate in the water. These organisms may be having disease vectors, thus increasing the fertilizer which content could proliferate the waterborne deceases.

Table 9.7: Fertilizers Consumption in India

	1951-52	*1993-1994*	*20010-2011*
Consumption (mil. Tonnes)	7.4	13.9	19.2
Increase (per cent) Since 1951	-	98.2	159.4

Source: Central Water Commission, 2000.

The use of large and unchecked pesticides in agriculture is another major cause of concern which greatly affects the quality of groundwater. Intensive cultivation of crops discharges chemicals from pesticides to seep into the groundwater, a process commonly known as leaching. Routine applications of pesticides for agriculture and indiscriminate disposal of industrial and domestic waste are increasingly being recognized as significant sources of water pollution. The high nitrate content in groundwater is mainly from irrigation run-off from agricultural fields where chemical fertilizers have been used indiscriminately. As indicated in the Table 9.6, since 1951 to 1994 the use of pesticides by Indian farmers increased by 249.84 per cent (Table 9.6). The associated agro food-processing industry is also a significant source of chemical as well as organic pollution in many of the regions of India. Apart from that India is likely to be afflicted by acid rain on a large scale. There were several parts of the country, where serious acidification of the soil as a result of acid rains was witnessed.

Table 9.8: Pesticides Consumption in India

	1951-52	*1993-1994*
Consumption (Tonnes)	24305	85030
Increase (per cent)	-	249.84

Source: Central Water Commission, 2000.

Right from the First Five Year Plan in 1951, the irrigation potential of India was only 6.5 million hectares, now it has increased up to 45 million hectares. But in the process the number of tube wells in India have increased up to 17 million. This number is quite high for any country or region of the world. Because of these numbers of tube wells, India is overexploiting its groundwater resources. In India, due to their political compulsion most of the State Governments are providing free electricity to the farmers for agricultural purpose. But it is not a rational decision as far as groundwater sustainability is concerned. Most of the farmers are utilizing groundwater beyond their requirement, because electricity is free. They keep their water tube well continuously on. Therefore, they are allowing the groundwater table to go down. Now we must understand what would be the effect when water level goes down. It is not merely that the water level goes down and people have to lift water from greater depth, it would lead to the ingress of saline water which would pose problem. It is estimated that till now more than 0.2 million hectare of land is now unutilizable due to water salinity.

This overuse of fresh water is not only being done by tube wells but also from the canal irrigation. In India, almost 0.3 million sq. km area is waterlogged and another 0.7 million km. would be waterlogged. In Punjab 25 per cent of the area is waterlogged. In Haryana also 25 per cent of the area is waterlogged. So on one hand we are **loosing** our productive land and also wasting our precious fresh water resources. It is clear that agriculture, is the single largest user of fresh water and therefore the major cause of degradation of surface and groundwater resources chemical **run-off and, has cause to be concerned about the water quality and ecosystem damage.**

Conclusion

Water is not only basic human need, but also is the basis of health, food security and economic development. But it has been clearly established that scarcity and contamination of fresh water resources is going to be a major environmental and socio-economic concern for India. Therefore, it is important for the planners to resolve the problem areas of India's fresh water resources. Still a large number of the population do not have any source of safe drinking water facility, most of the households of India particularly in urban centre are not connected with proper sewerage system.

Our waste water treatment capacity is insufficient and also overloaded, farmers are applying fertilizers, pesticides, irrigation without any guidelines. Most of the industrial units do not have waste water treatment plants and most importantly the demand for food is bound to increase in future.

Challenges are daunting, therefore, a comprehensive water conservation and management has to be adopted more rigorously. We have to revive our local level traditional water management systems. It is proved that the small and local water sheds utilize the water resources in controlled and effective manner than larger canal system. During monsoon a large amount of surface water remains unutilized and flows to the sea. Looking at India in a global context, the India has very low water storage capacity. For example, United States and Australia have capacity to store over 5,000 cubic meters for every citizen, China 2,500 cubic meters per capita and Morocco and South Africa 500 cubic meters per capita, India's storage capacity amounts to just 200 cubic meters per capita. It is quite obvious that if we enhance our water storage capacity it can solve our water scarcity problem. We also have to make people aware about the basic understanding and skills of water conservation and harvesting. The government should provide essential technical and financial support to the common households for rainwater harvesting. Finally, there is a dire need to increase our capacity to conserve the water resources at the national level and the Indian Government must balance competing demands between urban and rural, rich and poor, the economy and the environment.

REFERENCES

Central Ground (1996), *National Perspective Plan for Recharge to Ground Water by Utilising Water Board Surplus Monsoon Runoff*. CGWB Publication. Ministry of Water Resources.

FAO. Control of water pollution from agriculture Series title: FAO Irrigation and Drainage Papers—55 1996 (Agriculture).

http://www.lenntech.com/aquatic/introduction.

Julius, Q.S. and Bukey J.B. (1980), Assessment of Economic Contribution of Water Resources to National Development, Water Resource Institute of Civil Engineers, London.

Ministry of Environment and Forests, Government of India (2009): Report on Climate Change and India: Towards Preparation of a Comprehensive Climate Change Assessment.

OSSEC Business Network Switzerland Report (2010): Water treatment in India. Market Opportunities for Swiss companies, Stampfenbachstrasse 85, Zurich, Switzerland.

P.J.S. Bhamrah (1998). Non Judicious use of Water Resources Suggestions For Sustainable Management, Congres International De Kaslik—Liban—18-20 June, 1998.

Population Reports: The Coming Water Crisis, Volume XXVI, Number 1 September, 1998.

Prashant Gupta (2005). Underground Water Development in India—Trends, Crops, Kellog School of Management. www.gopio.net

Tyagi, S.K., J.S. Kamyotra, S.D. Makhijani, and S.P. Chakrabarthi (1998), "Water Resources Management and Pollution Assessment: Indian Scenario", *Journal IAEM*, Vol. 25: 15-21.

10

Participatory Sustainable Mangrove Conservation and Management of Coastal Wetlands

K. Arup Kumar Patro

Introduction

Global climate change is one of the greatest challenges that humans will face in this century. Although geological records show climatic changes throughout history, the present rate of global warming threatens the survival of entire ecosystems. Among the most at-risk ecosystems are mangroves, which are especially vulnerable to sea-level rise, but the good news is that not all coastlines with mangrove forests are projected to experience a rise in relative sea level.

About 40 per cent of the world's population residing within 100 km of the coastline. The major portions of coastal regions, particularly from the tropic, are characterized by ecological sensitive fragile ecosystems such as sea grass, corals, seaweeds, mangroves, obligate halophytic and sand dunes, essentially because they represent the interface between the land and the sea. Mangroves form one of these major habitats, sustaining various types of fauna and flora of socio-economic values. The physical structure of fishes and invertebrates is greatly influenced by mangroves.

Mangroves

Mangroves are woody trees or shrubs that grow in mangrove habitats or mangal (Hogarth, 1999). Mangroves grow only in

tropical and subtropical tidelands, which are frequently inundated with salt water, such as estuaries and marine shorelines. Mangrove trees form a specific ecological community, denominated as mangroves. Mangrove forests occupy about 17 million hectares of tropical coast worldwide: across Africa, Australia, Asia and America. The mangrove ecosystems provide protection against extreme weather events, such as storm winds and floods, as well as tsunamis. This is due to their capacity to absorb and dissipate the tidal surges that are associated with these events. They also contribute to the functioning of adjacent ecosystems, including terrestrial wetlands, peat swamps, salt marshes, sea grass beds and coral reefs.

Biodiversity

Rich in biodiversity, mangrove ecosystems provide a habitat for wide varieties of animal and plant species. Moreover, mangroves are home to many kinds of animal species due to the richness in food and form dynamic ecosystems. Live and decaying mangrove leaves and roots provide nutrients that nourish plankton, algae, shellfish, fish, crabs and shrimp.

Livelihoods

Furthermore, mangrove ecosystems provide many local people with basis of their livelihoods. Traditional economic activities vary from fishing and gathering of crustaceans to usages of the trees for timber or tannin production (*see* Fig. 10.1).

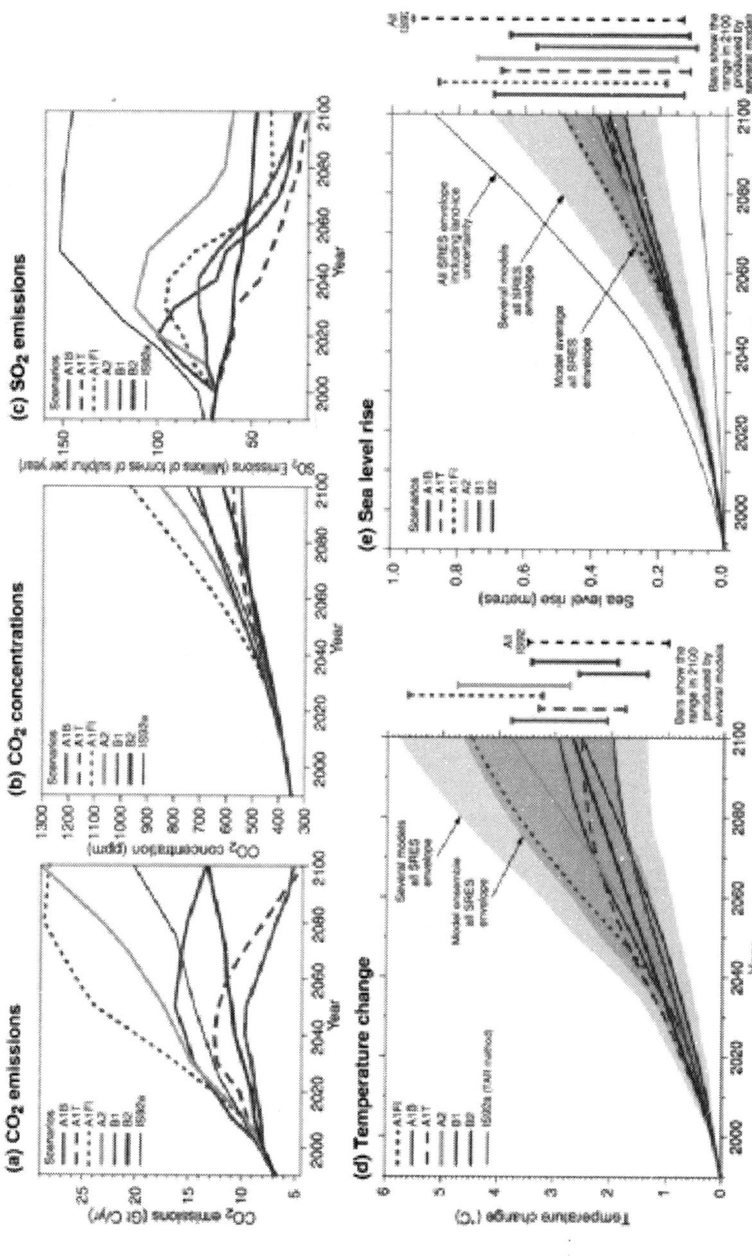

Fig. 10.1: The Global Climate of the 21st Century.

Potential Impacts of Climate Change

The global climate of the 21st century will depend on natural changes and the response of the climate system to human activities. Climate models project the response of many climate variables—such as increases in global surface temperature and sea level—to various scenarios of greenhouse gas and other human-related emissions. (a) shows the CO_2 emissions of the six illustrative SRES scenarios; (b) shows projected CO_2 concentrations; (c) shows anthropogenic SO_2 emissions; (d) and (e) show the projected temperature and sea level responses, respectively. (Copyright: Houghton *et al.*, 2001). Global climate change, specifically changes in temperature, CO_2, precipitation, hurricanes and storms, and sea level, combined with anthropogenic threats will threaten the resilience of mangroves.

Effects of Changes in Temperature

Since 1880, the Earth has warmed 0.6-0.8 °C and it is projected to warm 2-6 °C by 2100 mostly due to human activity (Houghton *et al.*, 2001). Mangroves are not expected to be adversely impacted by the projected increase in sea temperature (Field, 1995). Most mangroves produce maximal shoot density when mean air temperature rises to 25 °C and stop producing leaves when the mean air temperature drops below 15 °C (Hutchings and Saenger 1987).

Effects of Changes in CO_2

CO_2 significantly increases photosynthesis and the average growth rates in two mangrove species, *Rhizophora stylosa* and *Rhizophora apiculata,* but only when grown at lower salinity levels (Ball *et al.*, 1997).

Mangrove Adaptations that Help them Survive Sea-Level Rise

Mangroves have adapted special aerial roots, support roots, and buttresses to live in muddy, shifting, and saline conditions. Mangroves produce peat from decaying litter fall and root growth and by trapping sediment in the water. The process of building peat helps mangroves keep up with sea-level rise.

Effective Management

To encourage resilience to global climate change, mangroves need to be protected from anthropogenic threats, because mangroves that are healthy will also be better able to adapt to global changes. Over half of the mangrove areas are located within 25 km of urban centres inhabited by 1,00,000 or more people (Millennium Ecosystem Assessment, 2005). Close proximity to urban areas poses threats like pollution, dyking, channelization, unregulated felling, conversion to aquaculture or agriculture, and other forms of coastal development. Community based projects to reduce anthropogenic stress is important in conjunction with expanding our scientific understanding of the impacts of global change. Anthropogenic threats to mangroves are expected to increase with climate change.

Human Threats to Mangroves

Although mangrove ecosystems have tremendous value for coastal communities and associated species, they are being destroyed at alarming rates. Over the last 50 years, about one-third of the world's mangrove forests have been lost (Alongi, 2002). Human threats to mangroves include the overexploitation of forest resources by local communities, conversion into large scale development such as agriculture, forestry, salt extraction, urban development and infrastructure, and diversion of freshwater for irrigation (UNEP, 1994). The greatest human threat to mangroves is the establishment of shrimp aquaculture ponds.

Shrimp farms cover the area where mangrove forests once stood.

Benefits of Coastal Wetlands

Some of the primary benefits of coastal wetlands are flood protection and prevention of soil erosion. These ecosystems are on the front line when it comes to absorbing the brunt of storms. The forces which these wetlands encounter can be enormous as is the case with hurricane-strength storms. The vegetation of these wetlands helps anchor soils in place and also help to prevent soil erosion. However, these areas are facing serious threat due to the coastline erosion.

Threats to Coastal Wetlands

Development continues to remain a potent threat to coastal wetlands. Coastal areas are desirable properties especially with rising population levels. The only way some areas can increase the land available for development is by draining wetlands. Ironically, by doing so, they increase the vulnerability of these areas to additional land losses.

Why is the Coastal Programme Needed?

On the question as to why a coastal programme is required. It is worthwhile to mention that our Nation's coasts provide important fish and wildlife habitat, far beyond their limited geographic extent. Coastal ecosystems comprise less than 10 per cent of the Nation's land area, but support far greater proportions of our living resources.

Why are Wetlands Important?

There are many different kinds of wetlands and they all perform ecological functions, and produce certain goods and services that are valuable to humans such as:

Food Source: Food from the ocean is the world's chief animal protein source for all global citizens. Two thirds of all commercially caught fish spend their youth in coastal wetlands.

Economic Vitality: Bird watching, boating, fishing, beach combing, and other coastal activities attract national and international travellers.

Habitat: Coastal wetlands serve as breeding and nursery sites for oceanic species, birds and other wildlife.

Global Warming Management: A photosynthetic process that occurs in coastal wetlands removes carbon dioxide from the atmosphere and replaces it with oxygen.

Clean Water: Healthy, unabused wetlands are an effective, natural way to clean polluted waters and filter out sediments. Using wetlands as part of water treatment systems reduces cost and complexity.

Flood Protection: Storm water is absorbed by wetlands like a sponge. This protects surrounding areas from floods, preventing tragedy and expense.

Water Quality: Wetlands are one of the nature's most efficient water filters. Wetland plants and soils clean the water before it goes into groundwater or into rivers.

Erosion Control: Near shore, wetlands act as buffers to reduce shoreline erosion and stabilize banks.

Mangrove Forests in India

Mangrove forests are one of the most productive and bio-diverse wetlands on Earth. Mangroves are marine tidal forests and they are most luxuriant around the mouths of large rivers and in sheltered bays and are found mainly in tropical countries where annual rainfall is fairly high. Mangrove plants include trees, shrubs, ferns and palms. These plants are found in the tropics and sub-tropics on riverbanks and along coastlines, being unusually adapted to anaerobic conditions of both salt and fresh water environments. These plants have adapted to muddy, shifting, saline conditions. They produce stilt roots, which project above the mud and water in order to absorb oxygen. Mangrove plants form communities which help to stabilize banks and coastlines and become home to many types of animals.

Mangrove forests once covered font (three-fourth) of the coastlines of tropical and sub-tropical countries. Today less than 50 per cent remain, and of this remaining forest, over 50 per cent is degraded and not in good form. Many factors contribute to mangrove forest loss, including the charcoal and timber industries, urban growth pressures, and mounting pollution problems. However, one of the most recent and significant causes of mangrove forest loss in the past decade has been the consumer demand for luxury shrimp, or "prawns", and the corresponding expansion of destructive production methods of export-oriented industrial shrimp aquaculture. Vast tracts of mangrove forests have been cleared to make way for the establishment of coastal shrimp farm facilities.

Management of Mangroves in India

In India, mangroves occur on the West Coast, on the East Coast and on Andaman and Nicobar Islands, but in many places they are highly degraded. According to the Government of India (1987), India lost 40 per cent of its mangrove area in the last century. This chapter is an attempt to provide some considerations for conservation practitioners as they design conservation strategies for mangroves.

The project "Coastal Wetlands: Mangrove Conservation and Management" has been implemented with support from ICEF in four States, i.e. Tamil Nadu, Andhra Pradesh, Orissa and West Bengal to enhance national capacity and national action in the conservation and sustainable management of the mangrove wetlands of the east coast of India.

Results

The purpose of the project is to build capacity of the government agencies, involving local communities and also of the implementing agency to conserve, restore and sustain mangrove wetlands in an integrated manner through participatory approach. The outcome results of the same are mentioned below:

- Sustainable models for Joint Mangrove Management (JMM) were developed and information and knowledge on sustainable model for JMM was disseminated among the people.
- Appropriate institutional mechanisms are developed.
- Community and Forest Department are capable of jointly managing the mangrove wetlands.
- Different models developed and demonstrated for restoration and redevelopment of degraded mangrove wetlands.

- Alternatives are developed for fuel/fodder/timber/house-building materials that are currently being collected from the mangrove wetlands.
- Livelihood condition of the communities in Demonstration Villages improved.
- Scientific understanding developed, documented and disseminated amongst policy makers, Forest Department and communities.
- Appropriate models and policies advocated for adoption.
- Productive assets are created so that the resilience of the community to natural disasters may be strengthened.

Afforestation/Restoration of Degraded Mangroves and Conservation of Verdant Mangroves through Mangrove Management Units (MMUs)

The project attempted to conserve 746 ha. by constituting six Mangrove Management Units (MMUs) covering 10 villages. The project also undertook restoration of 240 ha. of degraded mangroves in Orissa (Mahanadi, Devi and Dhamara River mouth mangroves).

In addition to plantation, "Canal Technique" was adopted to regulate the salinity of the sites by periodical flushing of tidal water through the canals dug much before planting. 7 main canals and 22 feeder canals were dug covering an area of 512 ha.

Mangrove Nurseries

To overcome the low survival percentage, to generate alternative employment for the villagers and to reduce the transportation cost, the village level mangrove nurseries were established at Kandarapatia, Kajalpatia and Kalatunga of Mahanadi and Bandar and Naupal in Devi Mouth.

Management and Conservation

Luna Jungla Samrakshana Samiti (LJSS)

In each village, an institutional mechanism called Luna Jungla Samrakshana Samiti (LJSS) for joint restoration of degraded mangroves and conservation and management of verdant mangroves was established. The LJSS draws the micro plan and implements the activities which are approved by the resolutions. The MMUs were generally protected by *"Thengapalli"* and social fencing. Thengapalli is a unique method of protection of mangroves adopted, by which the villagers protect the forests by turn, by holding a *lathi* (stick).

Regional Forest Protection Committees (RFPCs)

Three RFPCs have also been formed in the Demonstration Villages (DVs) at Mahanadi, Devi Mouth and Dhamara Mouth to protect the mangroves collectively by associating representatives from Panchayats.

Homestead Plantations: Community and homestead plantations were raised in the demonstration villages which are helping in reducing the dependency on mangroves to meet their domestic needs.

Community Nurseries: 1,28,550 Multiple Purpose Tree (MPT) species were raised by the SHG members which were used for homestead plantations.

Land Based Alternatives: A Mangrove Dependency Survey shows that there is a reduction of 30 per cent dependency on mangroves for fuel, fodder and house building material in the demonstration villages. Plantation of multipurpose tree species has been undertaken in homestead as well as in common lands. The Baseline Survey Report of new demonstration villages shows that people are dependent on mangrove forests to a large extent for fuel, fodder and house building materials.

Inter-Nodal Bamboo Multiplication: Bamboo multiplication by vegetative propagation was taken up as one of the land based alternatives in the demonstration villages.

Fodder Cultivation: Fodder cultivation was demonstrated in Kharinasi-11 and Kalatunga by planting Para grass *(Brachina matica)*.

Socio-economic Activities: Smokeless Chullhas and Gas Stoves Distributed: 885 smokeless chullhas in eight (8) demonstration villages were distributed and fifteen (15) biogas plants in Kerabelari demonstration village were installed.

Self Help Groups

Mahanadi site 20 nos., Devi Mouth site 16 nos. and Dhamara Mouth 7 nos. of SHGs were formed. The matching grant has been provided to all the SHGs by MSSRF and two SHGs in Mahanadi site could also obtain bank loans.

Micro-enterprises: Fishing is the primary occupation of the community in all the demonstration villages. Due to obvious reasons, the fish catch as well as income is getting reduced day by day. To overcome this problem, different types of need-based micro-enterprises were undertaken.

On-Farm Activities

Distribution of Fruit Crop Seedlings: Horticultural crop seedlings like coconut, cashewnut and hybrid mango were distributed in the new demonstration villages for homestead plantations.

Distribution of K.B. Pump and LLHP and Storage Bin: The project provided KB pump and LLHP to facilitate micro irrigation system to distribute water to the kitchen garden for better yielding of crops. Winnower and paddy thresher were provided to the villagers for better agriculture practices.

Seed Distribution for Kitchen Garden: Vegetable seeds were distributed in the new demonstration villages to enhance kitchen gardens.

Biofertiliser: A small biofertiliser unit for production of Azalia has been developed in Kalatunga village to cater to the needs of the paddy growers.

Disease and Pest Management: Coconut trees of Naupal and Amarapat were severely affected by Opisinia arenosella and Rhinoceros beetle. Two different types of pest predators, Bracon hebetor and Bracon breviconis were used to control the pests.

Other Activities: Cultivation of Oyster mushroom was taken up. Linkages were established with government and developmental agencies for vermin composting.

Off-farm Activities: Some income generating activities were also introduced such as:

***Distribution of Poultry Chicks**:* Vanaraja, Ciriraja and Redrock were distributed in demonstration villages in keeping with the objectives of earning more income and providing occupation to the women in their leisure time. Galvanized storage bins have been distributed in DVs to keep the harvested crop safely from the attack of rat and storage pests.

Crab Fattening: Demonstration of crab fattening was started in Kalatunga village. A bamboo net (8' × 6') was placed in the creek and food supplied regularly.

Beehives: 20 beehives were distributed in 4 demonstration villages.

Information and Awareness

The following activities were undertaken and different types of IEC materials were used in different types of target groups.

Institutional Development and Capacity Building Activities

Nature Club (Eco Club): One Mangrove Eco-club was formed in Kalatunga for creating awareness of the school children towards the conservation and management of mangroves. Environmental Cultural programmes like street "Jhada" (the Cyclone) were also organized through the club of Kalatunga village.

Trainings: Most of the need-based trainings were conducted locally for enhancing their knowledge.

Policy dialogue: The PMC meetings were conducted for discussions related to Joint Mangrove Management. Draft guidelines were prepared and communicated to Ministry of Environment and Forests. State Guidance Committee meetings were held which provided a platform for interaction.

Community based Organization: Ten village level institutions were formed under the guidelines of JFM.

Strengthening the Capacity of FD: The project has considerably helped in strengthening the capacity of FD in the restoration of degraded mangroves.

Increased Participation and Empowerment of Target Groups: PRA was used to identify the needs and concerns of the VLIs, based on which the activities were implemented through a participatory approach.

Sustainability

Replication

The Fish bone type canal model developed in Pichavaram in Tamil Nadu and Andhra Pradesh has been successfully replicated in Orissa.

Resource Mobilization

Power tillers, Grain Bins and *Krishaka Bandhu* pump (eco-friendly) were bought with Govt. subsidies. NABARD, Nationalized Banks and Cooperative Banks provided loans to the Self Help Groups (SHGs).

Sustainability of the Local-Village Institutions

LJSS is collecting subscription from villagers for the institutional development. They also saved money from the project activities and contributed to VDF.

Micro-Enterprise Development

Most of the villages are dependent on fishing. Micro-enterprises were developed for alternative income.

Learning and Innovation

Case studies of success stories—Number of case study documented during the project period.

Innovation

- Fish Bone type technique canal for the restoration of degraded mangrove areas.
- Intern-nodal multiplication of bamboos.
- Provide alternate livelihood opportunities to the communities.
- Galvanised storage bins for better preservation of grains.
- Poultry varieties such as Giriraja, Vanaraja, Black rock, Kalinga Brown were distributed.

Technology Adoption

- Transferring the Fish bone type canal digging technology.
- Low cost multipurpose buildings were constructed with the help of the technology developed by UNDP. Rat trap technology.
- Smokeless *chullhas* developed by NEDCAP were introduced which were found more effective than the indigenous *chullhas*.
- Bersim and Para grass were introduced to meet the fodder needs of the cattle.

Documentation/Communication

The following documentation was generated during the project period:

- Atlas of Mangrove Wetlands of India Part 3—Mangroves of Orissa is in the press
- Mangrove Decade book
- Pamphlets
- National Workshop Proceedings
- Brochure in Oriya
- Training Manuals on Mangrove nursery raising
- Training Manual on Beehives
- A video documentary of the project interventions
- A newsletter which was published once could not be continued.

REFERENCES

Ahmad, E. (1972). *Coastal Geomorphology of India;* Orient Longman Ltd., Bombay.

Ambasht, R.S. and Ambasht, N.K. (1998). *Ecology of Indian Wetlands* [In: Ecology of Wetlands and Associated Systems; (Eds.): Majumdar, S.K.; Miller, E.W. and Brenner, F.J.]; the Pennsylvania Academy of Science; Easton.

Annual Report, 2002-2003 and 2003-2004 of M.S. Swaminathan, Research Foundation, Chennai.

Blasco, F. (1984). *"Climatic Factors and the Biology of Mangrove Plants"* [In: *The Mangrove Ecosystem: Research Methods* (Eds.): Snedakar, S.C. and Snedakar, J.G.]; UNESCO; Bungay, U.K.

Managing Mangroves for Resilience to Climate Change by Elizabeth McLeod and Rodney V. Salm, IUCN Resilience Science Group Working Paper Series—No 2.

Project End Evaluation Report, "Coastal Wetlands: Mangrove Conservation and Management (Devi Mouth, Mahanadi and Dhamara Mangroves)." April 2004.

11

Climate Change and its Impact on Local Indigenous Tribal Communities of Central India: A Case Study of Baiga Tribes of Madhya Pradesh

Manish Mishra

Introduction

The traditional societies, often referred to as indigenous or tribal people, have accumulated a whole lot of empirical knowledge on the basis of their experience dealing with nature and natural resources. In recent years, one of the concerns of environmentalists and those concerned with sustainable development has been to evaluate information and traditional knowledge (TK) of value derived from the natural environment in which traditional societies live. These benefits of ethno-biological significance are of three kinds: (a) economic—traditional crop varieties and lesser known plants and animals of food value, medicinal plants, etc. harvested from the wild; (b) ecologic/social—manipulation of biodiversity for coping with uncertainties in the environment and global change, for controlling soil water regimes and hydrology, soil fertility management through soil biological processes, and for efficient organic residue management; (c) ethical—cultural, spiritual, and religious belief systems centred around the concept of the sacred species, sacred groves, and the sacred landscapes (Ramakrishnan *et al.* (2001). Forests play a major role in climate mitigation strategies through carbon sequestration and the provision of products substituting fossil energy and materials. Furthermore, forests contribute significantly to regional climate regulation and to continuous water supply in large and small scale water cycles. These regulating services of forests including their alleviating functions can be essential for adaptation strategies by the local communities through climate change (Ravindranath *et al.,* 2006).

From the Batak peoples of Indonesia to the Karamojong in Africa, those who are least responsible for climate change are amongst the worst affected by it. They are often referred to in generic terms such as the world's poor or vulnerable groups, but these descriptions disguise the fact that specific communities often indigenous and minority peoples are valuable than others. They have close and unique relationship with nature and often the entire communities' livelihood depends on the environment. Indigenous people relationship with the environment is exceptional. They are the original inhabitants of their lands and consider themselves as guardians. These communities are already often poor and marginalized, are also more in danger when climate related disasters strike (Mihlar, F., 2008). Several indigenous communities make their living within vulnerable environments—in mountainous areas, in jungles or in dry areas and are thus often the first to discern and suffer the effects of climate change (Laub, 2008). Similar disruptions in livelihoods occur in other areas where indigenous peoples make their living—the glacier melts and forest fires impact their livelihoods. Even though they are not responsible for climate change, indigenous peoples are excessively affected by it. However, these peoples are not just victims of global warming; they have a critical role to play in supporting global adaptation to climate change. Several indigenous populations possess a unique knowledge of plant genetic diversity that maybe needed to fight plant and animal diseases, or they know how to breed varieties that can cope with the stressed environments.

Indigenous groups have always been confronted with changing environments. Their strategies for coping with the change have allowed them to successfully negotiate historical shifts in climate and environment, by modifying existing practice, shifting their resource bases or restructuring their relationships with the environment. There is growing awareness that indigenous peoples may find themselves not only on the frontlines of climate change impacts, but also of impacts due to the rapidly expanding efforts to mitigate climate change (Nakashima, 2008).

The Study Area

The district Mandla is situated in the east-central part of Madhya Pradesh. It lies in the catchments of river Narmada and its tributaries. Mandla has been richly endowed with dense forests. The world famous Kanha National Park is pride of Mandla and of the state. The district forms a part of Satpura hills, which separates the cotton growing of the south from the wheat growing extension of the Malwa Plateau on the north. The tropic of cancer thus passes through the north of the district. The total area of the district is 13,269 square km.

The Dindori district is situated at the eastern part of Madhya Pradesh touching Chhattisgarh state. It touches Shahdol in east, Mandla in west, Umaria in north, and Bilaspur district of Chattisgarh State in south. It is located at 81.34 degree longitude and 21.16 degree latitude. The holy river Narmada passes through the district. It is situated at a height of 1100 m above sea level amongst herbal-rich, Maikal mountain ranges. Dindori has many historical as well as spiritual places. The district is covered in seven blocks namely Dindori, Shahpura, Mehandwani, Amarpur, Bajag, Karanjiya and Samnapur. The Baiga are very particularly vulnerable tribal groups which can be found only in this district. The Baigas are known as the "National Human".

The Baiga tribals of Madhya Pradesh are the indigenous people who have lived in the forested areas of districts Mandla, Balaghat, Bilaspur and Rajnandgaon since time immemorial. It is said that the very first Baiga was born from the womb of Mother on the Hill of Elephants and he was named Nanga Baiga. Since then they have always been referred to as 'Sons and Daughters of Mother Earth'. The Baigas are the most primitive and interesting forest tribals of the district. Their few implements of agriculture are supplied by the local smith, and their food, supplemented by countless roots and fruits are obtained by their own slight excerptions. The Baiga takes coarse foods and shows no extravagance in this aspect. They eat coarse grain, *kodo,* and *kutki,* drink *pej,* eat little flour and are normally content with what little that they get. One of the prime foods is *pej* that can be made from grounding *macca* or from the water left from boiling rice. Local people gave testimony that this food is much more better and healthier than many other food that they eat. Also, beyond doubt they eat several items from the forest that includes primarily *Chirota Bhaji, Gular* leaves such as *chirota, chinch, chakora, sarroota, peepal,* etc. They also eat *Birar Kand, Kadukand* and other rhizomes. Mushroom is also a delicacy. Numerous fruits such as mango, char, jamun, tendu are also eaten. They hunt as well, primarily fish and small mammals

The great majority of the Baigas have now taken to plough cultivation, as perforce they must, seeing that *bewar* cultivation has been put a stop to everywhere both in *malguzari* and in Government forest, except the Baiga *chak* or Reservation. This is a block of some 36 square miles of rugged and inaccessible jungle in the heart of the Maikal Range, containing six (at present only four) villages, which was set apart for the Baigas about 35 years ago. Here they were allowed to pursue unfettered their ancestral methods of hunting and agriculture at the rate of one rupee per axe; but side by side with this concession attempts were made to wean them to plough cultivation. In fact, only about one-fifth of the Baiga population of the district live in the Baiga Reserve; and of these only 74 families, in only three of the six villages, now practice *bewar* at all. They limit their operations to an area of 292 acres, which with the necessary rotations represents a reservation for axe cultivation of 2,336 acres or one-tenth of the total area of the Reserve (*see* Fig. 11.1).

Results

Leaving aside few requirements consisting only of salt and the clothes they wear, their few implements of agriculture and their food, which is *kodokutki, Baiganitur* (pulse) and countless wild roots, tubers, fruits etc. the Baigas are least department on the outside world. They have little or no idea of the economy. However, they know their food requirements particularly in the stress seasons. Mostly they store wild tubers of various *Diocsorea* species and eat vegetables like potato in harsh seasons (Table 11.1). At the time of severe drought and natural calamities like flood, etc. they use these stored wild tubers to fulfil their food and nutritional requirements. Hence, they have their own strategy to cope up the extremes of the climate. On the other hand, few Baigas earn income after selling Tikhur (*Curcuma angustifolia*) powder and Baichandi (*Dioscorea hemona*) chips in the local haat bazaar. Most of the wild tubers are also used for medicinal purposes. They possess indigenous

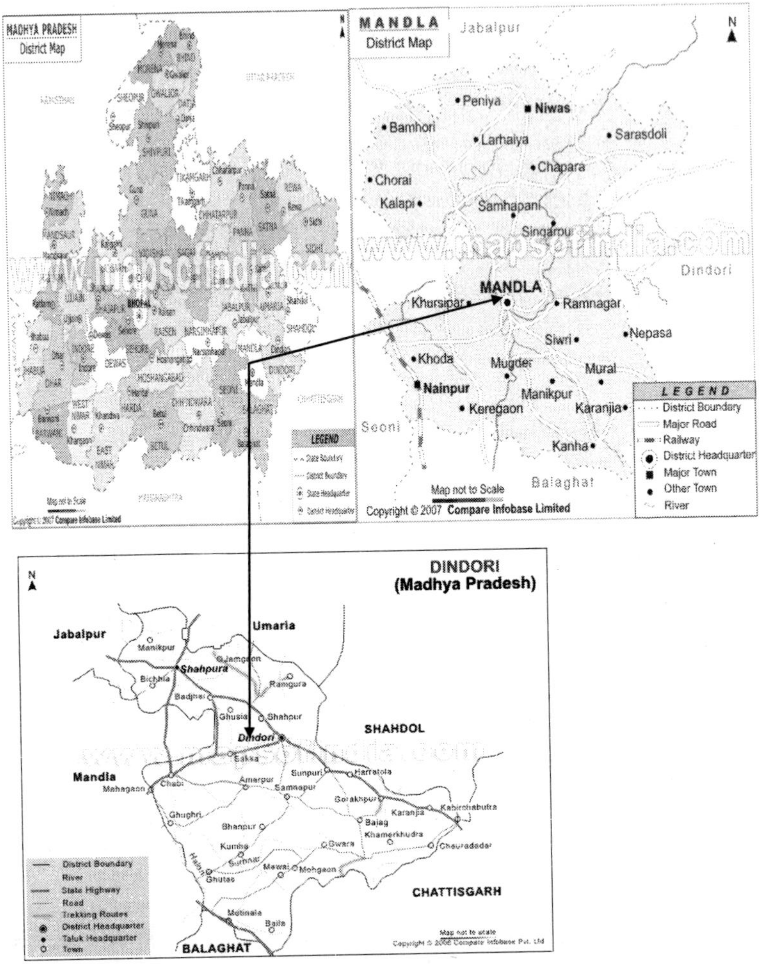

Fig. 11.1: Map showing Location of Madhya Pradesh and Mandla and Dindori District.

knowledge to use these wild tubers (*Dioscoreas*) in curing various ailments. Few of the tuberous crops collected from Mandla natural forests are listed below along with their food and traditional uses (Table 11.1).

Table 11.1: Wild Tuber Crops used as Food Supplement in Drought, Floods and Stress Conditions by the Baiga Tribes of Dindori and Mandla District of Madhya Pradesh

Name of tuber species	*Habit*	*Food/vegetable Use*	*Medicinal uses by Baiga Tribals*	*Time of collection*	*Circumstances*
Kanihakanda or Donchikanda (*Dioscorea bulbifera*)	Climber	After boiling used as vegetable	For disinfecting and cutting of Newly born child's Baby Nal	Jan.-March	At the time of Food shortage
Kadugeethkanda (*Dioscorea pentaphylla*)	Climber	After boiling and skin removal, used as vegetable		June	At the time of severe drought
Baichandikanda (*Dioscorea daemona*)	Climber	After thorough washing making Tuber chips, fried chips	Jaundice, Stomach ache, Malaria, Fever etc.	January	Selling chips is a major source of income
Lodangikanda	Climber	Used freshly (like potato)	no	June	At the time of severe drought
Saidukanda	Climber	After boiling edible (like potato)	no	June	At the time of water shortage and food shortage
Badainkanda	Herbaceous Climber	After boiling used as vegetable	no	Jan.-Feb.	At the time of vegetable shortage
Vidarikand Birarkanda or Bhui kumhra (*Prureria tuberosa*)	Perennial climber	Edible in unripe, raw stage, without boiling/processing	Diarrhea, bleeding, Swelling in body, potency etc.	November	At the time of drinking water shortage
Tikhurkanda (*Curcuma angustifolia*)	Herb	Tikhur paste and powder	Heat stroke, Weakness, Sexual strength	Nov.-Dec.	Selling powder is source of income
Ravikanda	Perennial climber	After boiling used as vegetable	Scorpion bite, poisonous insect bite etc.	May	At the time of drinking water shortage
Dhursikanda	Herb	After boiling used as vegetable	no	Dec.-Feb.	At the time of scarcity of vegetables

Traditional Agriculture Cropping Pattern in the Study Area

A Baiga cannot without offence lacerate the breast of mother Earth with the plough; and hence it is said they took to *Bewar,* their typical form of cultivation. A *Bewar* (shifting cultivation) consists of two to three acres of thick forest, often on a steep and almost precipitous slope. In the western literature it has been described as Slash and Burn type of agriculture. Baigas fell the trees in early January on the selected plots and there are trees left to dry. Before the first shower of the monsoons, the dry cuttings are set to fire. The burnt vegetation acts as thin productive layer on the earth. Usually 3-4 years the *Bewar* plots were cultivated and then left as fellow land from 6 to 10 years. On the break of the rains Kodon, Kutki, Baiganitur or sweet potatoes and other pulses crops were sown in the ashes without further preparation. At present most of the Baiga left this kind of agriculture due to restrictions from state governments and rules and regulations of land use.

Table 11.2: Agri-crops (*Bewar*) used by the Baigas of Dindŏri and Mandla District, Madhya Pradesh

S. No.	*Name of Crop (Baiga Name)*	*English Name*	*Scientific Name*
1.	Kutki	Kutki	*Picrorhiza kurrooa*
2.	Kodon	Kodo millets	*Paspalum scrobiculatum*
3.	Mandia	Finger Millets	*Eleusine coracana*
4.	Jowari	Sorghum	*Sorghum vulgare*
5.	Makkai	Maize	*Zea maize*
6.	Kulthi	Horsegram	*Macrotyloma uniflorum*
7.	Urda	Black gram	*Vign mungo*
8.	Rahar	Pigeon pea	*Cajanus cajan*
9.	Tili	Sesame	*Sesamum indicum*
10.	Kakdi	Cucumber	*Cucumis vulgaris*
11.	Semi	Kidney bean	*Phaseolus vulgaris*
12.	Jhujhru	Cow Pea	*Vigna unguiculata*
13.	Ravans	Adzuki bean	*Vigna angularis*

Baigas still believe that *Bewar* cultivation has several benefits over its adverse impacts on forest land and surrounding environment. Some of the benefits of traditional agriculture are:

(A) Soil Erosion: In the traditional agriculture, the tribals did not use plough during sowing. Crops were grown mostly in the slopes of the hill. After one year of cropping, they left the entire area for 2-3 years as fallow land. This duration is enough for increasing soil fertility and also helpful in checking soil degradation. However, there is no harmful impact on the crop production and fertility of soil due to *Bewar* cultivation and traditional methods of sowing adopted by the tribals.

(B) Food Diversity: *Bewar* is totally a mixed Agriculture system in which Tribals grow several varieties of crops, vegetables, pulses, in a limited area. They are getting all food and vegetables from the same land. However, for their food requirement they are dependent on market also.

(C) Food and Nutritional Security: In *Bewar* agriculture system, several crops were grown at a same time. The food crops which they grow (Table 11.2) are sufficient for their nutritional requirements. Few of the crops, i.e. *Kodo, Kutki* etc. are also used for medicinal purposes.

(D) Drought Resistance: In one crop system there are chances of crop failure due to floods, droughts and natural calamities, etc. Although, in the *Bewar* system during these calamities there are chances of crop production due to sowing of different crops and vegetables at a same time. The traditional pattern of crop selection (mixed) with other food crops and vegetable production is helpful in severe drought and flood conditions. Hence, they have strategy to cope up with the environmental change due to variation in the climate.

(E) Plant Biodiversity Conservation: Now experts are realizing that there is no harm in *Bewar* system of cultivation on forest and biodiversity. In this system, tribals left the area after 3-4 years of cultivation. After some time, new species of herbs, shrub trees and animals come up. The *Bewar* land, after few years again converts into forest land with complete biodiversity.

(F) Less Inputs and More Production: Due to very limited use of Agriculture implements and other farm machinery, the maintenance is free in the traditional agriculture system. There is also no expenditure on fertilizers, hybrid seeds and fungicides. The farmers are still using bio-fertilizers and bio-pesticides. On the other hand in one crop or modern system, huge inputs on these items are required to get more production.

(G) Prevent Economic Globalization Pressure on Agriculture: In the era of economic liberalization and globalization, there has been a lot of pressure on Agriculture sector. Several international companies are providing improved hybrid seeds, fertilizers, etc. to the farmers. On the other hand, organically produced agricultural products are also in great demand in the international markets. Due to water shortage and low fertility of the soil, the crop production is declining at a faster rate. *Bewar* cultivation and use of wild varieties of agricultural crops (shown in Table 11.2) helps Baigas to combat challenges of climate change.

The Baiga tribes of Baigachuk (Mandla and Dindori district) acquire a unique understanding of plant genetic diversity needed to fight plant and animal diseases and they know how to breed varieties that can cope up with the stressed environment. Local communities have always aimed to adopt variations in their climate. The Baigas living in and around the world famous, Kanha National Park possess knowledge about wild varieties of agricultural crops, wild edible tuberous plants (mostly *Dioscoreas*) and other vegetable crops. They eat several items from the forest that includes primarily *Chirota Bhaji, Gular* leaves, *Chirota, Chakora, peepal,* etc. They also eat *kanda* (tubers), mushrooms and numerous fruits such as mango, char, jamun, tendu, mahua, etc. They are growing various agri-crops using organic manure since time immemorial. Even after availability of hybrid seeds and improved varieties of agricultural crops, they are still growing Kodo (*Paspalum scrobiculatum*) and Kutki (*Picrorhiza kurrooa*) besides very low yield. Moreover, they grow all the vegetable agri-crops in their fields using organic and cow dung manure since time immemorial. Still they are using old traditional *grai* storage containers made of bamboo and cow dung with soil. Sometimes, they fill substantial amount of rice and maize in these containers and dump them inside the room after digging the ground surface. To protect from insect and pests, they always use Neem (*Azadirachta indica*) and *Nyctanthus* leaves. During the scarce period or drought year, they consume stored grain material efficiently. In this way, they are using traditional knowledge to curb the adverse impacts of climate change.

Due to continuous failure of monsoon, water shortage and low fertility of the soil, the crop production is declining at a faster rate. The farmers are facing acute shortage of fertilizers, insecticides and fungicides to protect their crops and also fail to increase production in the single crop system. The use of traditional seed and crop system is declining at a faster rate due to commercialization also. On the other hand, few Baiga communities of Dindori and Mandla district still practicing *Bewar* cultivation in their vicinity. Use of wild varieties of agricultural crops, mixed kind of crop cultivation and vegetable production helps the community in the drought conditions. They have number of wild crops and tubers, which are helpful in the scarce and extreme situations and serve as food material. It was also found that Baigas traditionally store wild tubers in their houses and are well prepared for any natural calamities. At the time of shortage of water, a large number of tuberous crops and varieties of seeds are available which can be grown under stressed conditions. Furthermore, they believe that the old system will return and they are prepared enough to cope up with the drought and changing climatic conditions.

Local communities have always aimed to adopt variations in their climate. To do so, they have made preparations based on their resources and knowledge accumulated through experience of past weather patterns. Local coping strategies are an important element of planning for adaptation. Climate change is leading communities to experience climate extremes more frequently, as well as new climate conditions and extremes. Traditional knowledge can help to provide efficient, appropriate and time tested ways of advising and enabling adaptations to climate change due to global warming. Several examples of local coping strategy are mentioned in the background papers of the workshop (UNCCD, 2004). In Asia, rural farmers have been practicing a range of agricultural techniques as coping strategies and tactics to enable sustainable food production and deal with the extreme events. These include intercropping and crop diversification such as introduction of wild varieties of seeds, traditional seeds. In Asia, farmers have traditionally observed a number of practices to adapt to climate variability for example—intercropping, mixed cropping, agro-forestry and developing new seeds varieties to cope with the local climate. In Latin America, local coping strategies include a variety of agricultural practices, ecosystem protection, and methods to adapt to extreme events. Farmers in Peru have been using an ancient irrigation and drainage system "Waru-Waru", or raised field agriculture, which makes it possible to bring production into low-lying, flood prone, poorly drained lands. The shallow canals provide moisture during droughts and drainage during the rainy season. The Aymaran indigenous people of Bolivia have been coping with droughts through the construction of small dams.

Conclusion

Baigas possess a unique knowledge of plant genetic diversity that maybe needed to fight plant and animal diseases and they know how to breed varieties that can cope with the stressed environment. They are among the first to undergo harsh and erratic climatic conditions, and a generalized lack of empowerment to claim goods and services to which other population groups have greater access. Increasingly tough climatic conditions and limited rights to land and other basic resources risk jeopardizing the lives and livelihoods of Baigas that hold the key to long term continued existence. Though, the Baigas are not just victims of global warming; they also have a critical role to play in supporting global adaptation to the climate change. The tribal communities, i.e. Baigas are the custodians of unique knowledge and skills and the genetic and biological diversity in plant and animals that may be vital in adapting to the climate change.

Community participation can work wonders. Attempts are being made by way of peoples committees to manage natural resources in more and more eco friendly manner. Experiences and practices in community based climate change mitigation shows the positive impact of the participatory approach to climate change. Involvement and participation of the local communities in climate change reduction programmes receives the highest priority in the present approach as they are affected by the global warming, climate change, etc. and more importantly, they are the first responders to a climate change event. In absence of any specialized skills, they rely on traditional coping and survival mechanisms to face and respond to the event before they start receiving any outside help. Preparedness measures such as the maintenance of inventories of resources and the training of local peoples/ community stakeholder to manage climate change and global warming, etc. are vital components to mitigate climate change. Furthermore, this should be an ongoing, regular function of all the Government departments at all levels, in active collaboration with local/indigenous people.

REFERENCES

Biswas, N. (2007). *Bewar Swaraj: Traditional science of natural organic farming.* Publ. by Nirman Vaikalpik Vikas evam Sahbaghi Shodh Sansthan. Sijhoura, Bhua-Bichhiya, Mandla (MP), pp. 1-106.

Laub, Regina. (2008). "NWFPS and climate change." *Non Wood News.* No. 17, July, 2008, pp. 7-8. FAO, Rome, Italy.

Mihlar, Farah (2008). Voices must be heard: minorities and indigenous people combating climate change. Briefing note. Minority Rights Group International. London, United Kingdom, pp. 1-12.

Nakashima, Douglas (2008). An Indigenous Knowledge Forum on Climate Change Impacts. Local and Indigenous Knowledge System (LINKS). UNESCO, 2008. Website-http://portal.unesco.org/science/en/ev.php-Url_id=6550&url_do=do_topic&url_section=201.html

Ramakrishnan P.S., Neil J. Smelser, and Paul B. Baltes (2001). *Ethnobiology International Encyclopedia of the Social and Behavioural Sciences.* Oxford, UK: Pergamon, pp. 4846-4852.

Ravindranath, NH., Joshi, N.V., Sukumar, R. and Saxena, A. (2006). "Impact of climate change on forests in India." *Current Science,* Vol. 90, No. 3, pp. 354-361.

TERI (2004). Impacts of Climate Change. Report of Tata Energy Research Institute, New Delhi. http://www.edugreen.teri.res.in/explore/climate/impact.htm.

UNCCD (2004). *United Nations Convention to Combat Desertification.* In Climate Change. Viterbo, Italy, 5-7 April, 2004. www.unccd.int/workshop/menu.php-21k.

12

Ocean Acidification

Great Risk to Oceans by Climate Change

H.M. Parekh and *P.U. Poriya*

Introduction

Global change encompasses a multitude of environmental and ecological changes that have been observed during the last decades as a direct or indirect result of increasing human population, the burning of fossil fuels and consequent accumulation of greenhouse gases in the atmosphere. Changes in the climate are the most prominent manifestation of global change. Climate change impacts the ocean through generalized temperature rise, shifts in wind and radiation regimes, changes in the hydrological cycle, alterations related to oceanic circulation and stratification, changes in the frequency of episodic extreme events (such as storms) and acidification. The most immediate and direct effect of global change at a global ocean scale will be, however, the increment in the surface seawater temperature, down to tens of meters. Climatic models predict that water temperatures will increase by a few degrees during this century (Timmermann *et al.*, 1999; Meehl *et al.*, 2007). These predictions have been confirmed by long-term observations in European seas where sea surface temperature rates of increase have been around 0.01°C yr^{-1} since the 1860s (Wiltshire & Manly, 2004; Vargas-Yañez *et al.*, 2005; Mackensie & Schiedek, 2007). It is likely that such changes will deeply affect different aspects of the structure and functioning of marine ecosystems (Edwards & Richardson, 2004; Wiltshire *et al.*, 2008; Montoya & Raffaelli, 2010).

Climate change is an important factor in determining the past and future distributions of biodiversity (Peterson *et al.*, 2002;

Parmesan and Yohe, 2003; Root *et al.*, 2003; Thomas *et al.*, 2004a, b; Rosenzweig *et al.*, 2008). In the ocean, the pattern of marine species richness, notably for fish and invertebrates, is strongly related to environmental factors (Macpherson 2002). In addition, observations and theory suggest that marine species respond to ocean warming by shifting their latitudinal range (Perry *et al.*, 2005; Parmesan, 2006; Hiddink and Hofstede, 2008; Mueter & Litzow, 2008) and depth range (Dulvy *et al.*, 2008). Such species responses may lead to local extinction and invasions, resulting in changes in the pattern of marine species richness. Overall, changes in pattern of species richness may disrupt marine biodiversity and ecosystems, and impact commercial fisheries (Roessig *et al.*, 2004; Worm *et al.*, 2006; Chesung *et al.*, 2008).

When carbon dioxide (CO_2) is absorbed by seawater, carbonic acid is formed, reducing the water's pH level and the concentration of carbonate ion. This process is commonly called ocean acidification. Recent studies suggest that the oceans are becoming less able to absorb as much carbon dioxide as they have in the past, decreasing their ability to buffer against climate change. By absorbing massive amounts of carbon dioxide, the pH (a measure of acidity) in ocean surface waters has dropped approximately 30 per cent. Naturally, the oceans are alkaline, meaning they are above seven on the pH scale. Prior to industrialization, the pH of the oceans was on average about 8.2; however, since then, the average pH has dropped about 0.1 units. While this may seem like a small change, it actually represents a 26 per cent increase in acidity.

By now, most of us know that carbon dioxide-induced global warming is causing changes in the ocean temperatures and precipitating a rise in the sea levels. Perhaps surprisingly, ocean acidification is considered to be one of the most serious consequences of increased CO_2 in the atmosphere.

It is estimated that the ocean has absorbed more than 528 billion tons of CO_2 from the atmosphere—about one-third of human-caused carbon emissions—since the beginning of the industrial revolution.

On the one hand, the ocean's absorption of CO_2 helps reduce the amount of greenhouse gases in the atmosphere—a positive, useful function in these dire times. On the other hand, the ocean's absorption of CO_2 and the ensuing drop in seawater pH level has widespread and devastating effects on marine and human life. The lower pH level inhibits the ability of many marine plants and animals to build their shells and skeletal structure, in some cases even dissolving the shells. Ocean acidification is particularly harmful to surface and deep-water corals, plankton, snails, lobsters, clams, oysters and other molluscs.

Beyond individual organisms, ocean acidification holds broad repercussions for entire ecosystems—and human survival. The plant and animal organisms most affected by ocean acidification provide critical habitat and food sources for other organisms. Ocean acidification exacerbates problems already faced by fisheries and marine ecosystems, including over-fishing, pollution, excess nutrients, invasive species, and habitat destruction.

Carbon emissions from human activity have led to a 30 per cent increase in the acidity of ocean surface waters. Scientists and researchers estimate that seawater pH will drop rapidly to 7.7 from a normal 8.1. In June 1970, the world's science academies came together in Bonn, Germany, to warn of dire consequences for food production and the livelihoods of millions if the issue of ocean acidification is not addressed. Their joint statement asserts that oceans are more acidic now than they have been for 8,00,000 years. "The effects will be seen worldwide, threatening food security, reducing coastal protection and damaging the local economies that may be least able to tolerate it", Martin Rees, Pers comment.

Ocean Acidification: Impact on Marine Life

Corals, calcareous phytoplankton, mussels, snails, sea urchins and other marine organisms use calcium (Ca) and carbonate (CO_3) in seawater to construct their calcium carbonate ($CaCO_3$) shells or skeletons. As the pH decreases, carbonate becomes less available, which makes it more difficult for organisms to secrete $CaCO_3$ to form their skeletal material. For animals in general, including invertebrates and some fish, CO_2 accumulation and lowered pH may result in acidosis, or a build up of carbonic acid in the organism's body fluids. This can lead to lowered immune response, metabolic depression, and behavioural depression affecting physical activity and reproduction, and asphyxiation. Since the oceans have never experienced such a rapid acidification, it is not clear if ecosystems have the ability to adapt to these changes. Effects of ocean acidification on organisms and ecosystems are still poorly understood. Over the last few years, research has intensified significantly to fill the many knowledge gaps. Studies of marine calcifiers indicate that most but not all of them exhibit reduced calcification with increased ocean acidification (Fabry *et al.*, 2008). Marine calcifiers differ because they use various mechanisms to control their internal micro-environments where calcification takes place. In addition, different life stages of marine calcifiers respond differently.

Effect of Ocean Acidification on Phytoplankton and Zooplankton

Changes in the carbonate chemistry of the ocean may have a strong negative impact on many phytoplankton and zooplankton species that form the base of the marine food chain. In almost all calcifying organisms tested, ranging from single-celled organisms up to reef building corals, there is a decrease in the ability of the organism to produce calcium carbonate in more acidic waters. One study has documented the changes in two species of coccolithophores grown under conditions expected by the end of this century, where both species show significant decreases (25-45 %) in calcification rates and clear signs of structural damage in their shells, which may affect their physical functioning and reproduction. However, not all species of calcifying organisms are negatively affected by increased acidity, and more research is needed to understand these mechanisms and possible adaptation pathways.

Pteropods are small "winged snails" that form the basis of the food chain for many commercial fish species. A recent study has observed shell dissolution in living pteropods when exposed to the carbonate content of the ocean expected in the next 50 years in the high latitudes. If ocean acidification leads to disturbances in the populations of these organisms, other non-calcifying organisms may out-compete them for food and nutrients, leading to a change in the ecosystem composition of the system. Although some of the affected organisms are important prey for higher organisms, it is not yet clear how such changes would affect fisheries.

Corals

Nearly 500 million people depend on healthy coral reefs for sustenance, coastal protection, renewable resources, and tourism, with an estimated 30 million of the world's poorest people depending entirely on the reefs for food. Coral reefs face two challenges from increasing atmospheric CO_2. First, higher CO_2 concentrations in the atmosphere are linked to warmer global temperatures, which in turn lead to warmer water temperatures. Corals are very sensitive to temperature change: a 1-2 °C change in local temperature above their normal summer maximum can lead to a phenomenon called 'bleaching', whereby the corals expel their vital algal symbionts (algae which live in the cells of the coral), leaving the coral tissues translucent.

In 1998, a single bleaching event led to the loss of almost 20 per cent of the world's living coral. Corals can recover from these events but repeated episodes are likely to weaken the coral ecosystem, making them more susceptible to disease and causing a loss of biodiversity. The second challenge faced by corals is the increasing acidity of the water caused by higher CO_2 concentrations. Lowered calcification rates affect the reef's ability to grow its carbonate skeleton, leading to slower growth of the reef and a more fragile structural support, which makes it more vulnerable to erosion. By the middle of this century, the estimated reduction in calcification rates may lead to more reef area erosion than can be rebuilt through new calcification.

Other Marine Animals

The rate at which mussels and oysters build their shells is also lower in a high CO_2 (low pH) environment, which could have direct consequences on the aquaculture of molluscs, an industry of 12 million tonnes per year and a market value of over $10.5 billion US dollars. Higher marine life forms, animals such as invertebrates and even fish may be affected by lower pH environments through acidosis (an increase in carbonic acid in the body fluids causing lower pH values in blood) leading to lowered resistance, metabolic depression, behavioural depression affecting physical activity and reproduction, and asphyxiation.

Cephalopods such as squid seem to be particularly sensitive

to CO_2 increases because their energy-demanding way of swimming requires a good supply of oxygen to the blood, which is impaired by lowered blood pH values. Depending on the effect of other stressors like warmer temperatures, sensitivities may differ between life stages of a species, with larvae possibly being more sensitive. The full range of physiological mechanisms eliciting completely animal responses to elevated CO_2 levels is still insufficiently explored. Current experiments address the extent to which genomic information defines sensitivity and if animals can physiologically adapt to elevated CO_2 levels. Direct effect of CO_2 on marine mammals (seals and whales) or birds are not expected because they breathe air, and thus will not be directly affected by acidification of the surrounding seawater. Changing food webs will, however, affect these animals and their well-being in ways that are not fully understood.

According to a report on ocean acidification by the Royal Society (Raven *et al.*, 2005):

- Polar and sub-polar surface waters and the Southern Ocean are projected to be aragonite undersaturated by 2100, and Arctic waters will be similarly threatened.
- Ocean acidification will likely threaten ecosystem dynamics in the areas where marine calcifiers play dominant roles in the food web and in carbon cycling.
- Lower pH in the oceans may inhibit the ability of echinoderms, such as sea urchins and molluscs, to form skeletal material.
- Ocean acidification could have large consequences for benthic (bottom-dwelling) organisms, including in particular benthic calcifiers, by changing the pH levels sediment communities are exposed to and by causing changes in the biological pump.
- Some unicellular species that form the base of marine food chains could experience reduced growth and fitness as a result of decreased pH levels.
- The overall reaction of marine biological carbon cycling and ecosystems to a warm and high-CO_2 world is not yet well understood, but there is a risk that a decrease in calcifier productivity could lead to cascading effects throughout the food chain.

What Regions of the Ocean will be Affected First?

The first areas to be impacted by ocean acidification are high-latitude regions and deep water areas, where the natural carbonate levels are lowest and closest to becoming undersaturated with respect to carbonate. *Cold water coral communities*, which often occur at depths of hundreds of meters and are found in all the world's oceans, are slow-growing and fragile ecosystems that are thought to be important nursery habitats for many commercially important fish species as well as home to thousands of other organisms, making them hotspots of biodiversity that are still poorly understood. Other *benthic ecosystems and organisms* such as molluscs, starfish, and sea urchins are also expected to be impacted early owing to the already-low carbonate concentrations in the deep-sea environment.

As acidification continues and the carbonate saturation horizons becomes shallower, it is estimated that the *surface waters of high latitude regions* will become undersaturated with respect to aragonite by 2050, with calcite saturation levels becoming undersaturated 50-100 years later. This will affect all calcareous organisms in the productive surface waters, including coccolithophores, foraminifera, and pteropods, which form the base of the marine food-web for many species and play an important role in the natural cycling of carbon from the surface to deep ocean. Effects first seen in the high-latitude surface waters will also be seen in the *surface waters of other regions* as CO_2 levels increase, with impacts determined by the composition and functioning of each specific ecosystem.

Coral reef ecosystems in the warm waters of tropical and subtropical regions are among the most diverse and productive of marine ecosystems, supporting hundreds of millions of people directly through subsistence harvesting. These valuable ecosystems, which are found mostly in developing countries, will experience reduced calcification rates as ocean acidification progresses, leading to slower growth rates and a more fragile skeletal structure. It is estimated that by 2050, we may be losing more reef area to erosion than can be rebuilt through new calcification. There is minimal understanding of how the combined effects of warming seas (which cause coral bleaching) and reduced calcification due to acidification will interact.

Ocean Acidification: Effect on Societies and Economies

Ocean acidification may trigger a chain reaction of impacts through the marine food web that will affect the multi-billion dollar commercial fisheries and shellfish industries, as well as threatening the food security for millions of the world's poorest people. Larval fish and shellfish may be especially vulnerable. Ocean acidification may render most regions of the ocean inhospitable to coral reefs, affecting tourism, food security, shoreline protection, and biodiversity. Coral reefs may be particularly affected because of the combined impact of coral bleaching caused by increased water temperatures and ocean acidification. The ocean's capacity to absorb CO_2 from the atmosphere is being degraded by ocean acidification, which will make it more difficult to stabilize atmospheric CO_2 concentrations. The cost of stabilizing atmospheric CO_2 at a level that will avoid most of the negative impacts is lower than the cost of inaction, and stabilization is achievable with technology that can be deployed now and in the near future.

Thus, it can be concluded that global warming is a major issue and its main cause is release of CO_2 primarily through the burning of the fossil fuels and biofuels. Over the years, the oceans have helped this problem by absorbing excess carbon dioxide. But the rate at which ocean acidification is increasing this is leading to an alarming situation. While the existence of ocean acidification is now known among scientists, including oceanographers, the public, and even many policy-makers are still largely unaware of it. Raising awareness about the important role of oceans in general to address climate change and other environmental issues is crucial and could be undertaken immediately by the international community.

REFERENCES

Cheung, W.W.L., Lam, V.W.Y. and Pauly, D. (2008a). Modelling Present and Climate-Shifted Distribution of Marine Fishes and Invertebrates. Fisheries Centre Research Report 16(3), University of British Columbia Vancouver.

Cicerone, R. *et al.*, (2004). "The Ocean in a High CO_2 World", *Oceanography,* 17, 72-78.

Dulvy, N.K., Rogers, S.I., Jennings, S., Vanessa, S., Dye, S.R. and Skjoldal, H.R. (2008). "Climate change and deepening of the North Sea fish assemblage: a biotic indicator of warming seas." *Journal of Applied Ecology,* 45, 1029-1039.

Edwards, M., Richardson A.J. (2004) "Impact of climate change on marine pelagic phenology and trophic mismatch." *Nature,* 430, 881-884.

Feely, R.A., *et al.* (2004). "Impact of Anthropogenic CO_2 on the $CaCO_3$ System in the Oceans", *Science,* V 305, 362-366.

Gattuso, J.P., Allemand D. and Frankignoulle M. (1999). "Photosynthesis and calcification at cellular, organismal and community levels in coral reefs: a review on interactions and control by carbonate chemistry." *American Zoologist,* 39(1):160-183.

Gazeau, F., Quiblier C., Jansen J.M., Gattuso J.-P., Middelburg J.J. and Heip C.H.R., 2007. "Impact of elevated CO_2 on shellfish calcification." *Geophysical Research Letters,* 34, L07603.

Hiddink, J.G. and Hofstede, R.T. (2008). "Climate induced increases in species richness of marine fishes." *Global Change Biology,* 14, 453-460.

Hoegh-Guldberg, O. (2005). "Low coral cover in a high-CO_2 World", *Journal of Geophysical Research,* 110, C09S06, doi: 10.1029/2004JC002528.

Hood, M. (2004). "A Carbon Sink that can no longer cope?", *UNESCO's World of Science,* Vol. 2(4): 4, 2-5.

Mackensie B.R., Schiedek D. (2007). "Daily ocean monitoring since the 1860s shows record warming of northern European seas." *Glob. Change Biol.* 13, 1335-1347.

Macpherson, E. (2002). Large-scale species-richness gradients in the Atlantic Ocean. *Proceedings of the Royal Society of London Series* B 269, 1715-1720.

Meehl, G.A., *et al.,* (2007). *The physical science basis. In Climate change 2007 IPCC Report,* pp. 748-845. Cambridge, UK: Cambridge University Press.

Montoya, J.M., Raffaelli D. (2010). "Climate change, biotic interactions and ecosystem services." *Phil. Trans. R. Soc.* B 365, 2013-2018.

Mueter, F.J. and Litzow, M.A. (2008). "Sea ice retreat alters the biogeography of the Bering Sea continental shelf." *Ecological Applications,* 18(2), 309-320.

Parmesan, C. (2006). "Ecological and evolutionary responses to recent climate change." *Annual Review of Ecology, Evolution and Systematics,* 37, 637-669.

Parmesan, C. and Yohe, G. (2003). "A globally coherent fingerprint of climate change impacts across natural systems." *Nature,* 421, 37-42.

Perry, A.L., Low, P.J., Ellis, J.R. and Reynolds, J.D. (2005). "Climate change and distribution shifts in marine fishes." *Science,* 308, 1912 1915.

Peterson, A.T., Ortega-Huerta, M.A., Bartley, J., Sa'nchez-Cordero, V., Sobero'n, J., Buddemeier, R.H. and Stockwell, D.R.B. (2002). "Future Projections for Mexican faunas under global climate change scenarios." *Nature,* 416, 626-629.

Poertner, H.O., M. Langenbuch, and B. Michaelidis (2005). "Synergistic effects of temperature extremes, hypoxia, and increases in CO_2 on marine animals: From earth history to global change." *Journal of Geophysical Research—Oceans, 110,* C09S10, doi: 10.1029/2004JC002561.

Riebesell *et al.* (2000). "Reduced calcification of marine phytoplankton in response to increased atmopsheric CO_2", *Nature,* 407, 364-367.

Roessig, J.M., Woodley, C.M., Cech, J.J. and Hansen, L.J. (2004). "Effects of global climate change on marine and estuarine fishes and fisheries." *Reviews in Fish Biology and Fisheries,* 14, 251-275.

Root, T.L., Price, J.T., Hall, K.R., Schneider, S.H., Roxenzweiq, C. and Pounds, J.A. (2003). "Fingerprints of global warming on wild animals and plants." *Nature,* 421, 57-60.

Rosenzweig, C., Karoly, D., Vicarelli, M. *et al.,* (2008). "Attributing physical and biological impacts to anthropogenic climate change." *Nature,* 453, 353-357.

Status of Coral Reefs of the World (2004). ed. C. Wilkinson, Australian Institute of Marine Science.

Thomas, C.D., Cameron, A., Green, R.E. *et al.,* (2004a). "Extinction risk from climate change." *Nature,* 427, 145-148.

Thomas, C.D., Williams, S.E., Cameron, A. *et al.* (2004b). "Biodiversity conservation: uncertainty in predictions of extinction risk/effects of changes in climate and land use/climate change and extinction risk (reply)." *Nature,* 430, 1-2.

Timmermann A., Oberhuber J., Bacher A., Esch M., Latif M., Roeckner E. (1999). Increased El Nino frequency in a climate model forced by future greenhouse warming. *Nature,* 398, 694-697.

Vargas-Yañez M., Salat J., Fernandez de Puelles M.L., Lopez-Jurado J.L., Pascual J., Ramirez T., Cortés D., Franco I. (2005). Trends and time varibility in the Northern continental shelf of the western Mediterranean. *J. Geophys. Res.* 110, 1-18.

Wiltshire K.H., Malzahn A.M., Wirtz K., Greve W., Janisch S., Mangelsdorf P., Manly B.F.J., Boersma M. (2008). Resilience of North Sea phytoplankton spring bloom dynamics: an analysis of long-term data at Helgoland Roads. *Limnol. Oceanogr.,* 53, 1294-1302.

Wiltshire K.H., Manly B.F.J. (2004). The warming trend at Helgoland Roads, North Sea: phytoplankton response. *Helgoland Mar. Res.* 58, 269-273.

Worm, B., Barbier, E.B., Beaumont, N. *et al.* (2006). "Impacts of biodiversity loss on ocean ecosystem services." *Science,* 314, 787-790.

13

Climate Change

Issues and Challenges

Vinod K. Sharma

Introduction

On February 2, 2007, the United Nations Scientific panel studying Climate Change declared that the evidence of a warming trend is "unequivocal", and that human activity has "very likely" been the driving force in that change over the last 50 years. This report of IPCC was more conclusive as compared to the report released in 2001 which only pointed to the likeliness of human activity being responsible for climate change. The fact that climate change is real, that it is happening and that its impact is devastating millions is no longer news. Research by renowned scientists worldwide has established the fact that the temperature of the earth has increased by about 0.7 degree Celsius. It is also a generally agreed fact that the increase in temperature is due to the concentration of Green House Gases in the atmosphere including Carbon Dioxide which today stands at 380 ppm. Due to this concentration of Carbon Dioxide in the atmosphere the earth has reached a climatic "tipping point" since a further increase in Carbon Dioxide levels will obviously increase the heating of the earth which would bring in its wake faster melting of ice caps, rising sea levels, more frequent heat waves, water shortages for up to 250 million people and the risk of extinction for up to 30 per cent of all species. These are serious consequences which in addition to causing widespread loss and destruction would also seriously hamper the fulfilment of the Millennium Development Goals. It is imperative that global warming should be restricted to the extent possible collaborative action between nations.

Although the science of climate change has been accepted, its politics is still contested with developed nations refusing to take action on reduction of Green House Gases unless the developing ones do the same. The developing nations cite their right to develop as an excuse for inaction and continue to argue that the developed nations, which are the biggest emitters, should curb their emissions first. While precious time is lost in such arguments, Carbon Dioxide levels are increasing continuously and developing nations are continuously moving up on the list of emitters. Today America, China, Indonesia and India, in that order, are considered to be the four largest emitters. In this ongoing dispute between developing and developed nations a fact that has been largely forgotten is that the impacts of climate change have spared neither the developed nor the developing nations. The climate change events that have occurred in various parts of the world bear testimony to this fact. It would be interesting to note that most of these impacts have occurred in the last decade and have been increasing in frequency and intensity as predicted by scientists.

Indian Scenario

India has also not been spared by the impact of climate change. The country which is home to 1028.7 million people has a high population density of 344 persons per sq. km. which makes most areas of the country extremely vulnerable to various kinds of disasters which are bound to escalate with climate change. The country also has a long (7500 km) and populous coastline with a higher population density of 455 persons per sq. km. This coastal population depends heavily on sea based livelihood sources and is highly vulnerable to various disasters originating in the sea. Such sea based disasters have only worsened in the last decade. Another major threat that climate change poses to the country is through its impact on agriculture. Since agriculture is a source of livelihood for 65 per cent of the population and contributes to 27 per cent of the GDP, adverse impact on Agriculture would have serious effect on the economy of the country. In addition, it will also have an impact on food security and health. It has been estimated that 0.5 degree rise in temperature would result in reduction of wheat yields by 10 per cent. However, the loss of crops and agricultural land due to disaster intensification and land degradation caused by climate change cannot be estimated. The extreme weather events that have occurred in the country in the last decade show that various parts of the country are already reeling under the impacts of climate change.

- In the North, Himalayan Glaciers are receding by about 16m every year. The receding glaciers have major implications for water availability in the glacier fed rivers such as Ganga and Yamuna which serve as the main source of water supply to settlements along their banks. An estimated 750 million people live in the watershed areas of rivers originating from Himalayan glaciers.
- An Action Aid study has shown that water levels in almost all streams in Kashmir valley have fallen by 2/3rd and Groundwater levels have dropped by more than 1/3rd. These problems are occurring mainly due to the receding glaciers.
- In May 2002, severe heat waves conditions were recorded in coastal districts of Andhra Pradesh. Earlier, the heat waves were restricted to the interior areas of the state.
- In 2003, twenty two out of twenty three districts in Andhra Pradesh were in the grip of a heat wave. In the same year, more than 80 per cent of the state faced a drought.
- In 1999, two cyclones hit the state of Orissa in quick succession. About 15 million people were affected.
- In the Sunderbans delta for islands, Bedford, Lochahara, Suparibhanga and Kabasgadi, have submerged due to sea level rise rendering 6000 families homeless.
- The desert region of Barmer, in Rajasthan was affected by floods in 2006. The floods claimed the lives of 140 people.
- The city of Mumbai was flooded due to heavy rains in 2005. 944 mm rainfall was recorded in a period of 24 hours. About 1000 people killed in the floods.

The major impacts of Climate Change on various regions of India are as under:

East

The increased disaster frequency and intensity indicates that the state is under severe stress due to climate change impacts. In the past decade, the state has experienced various disasters due to extreme weather conditions such as heat waves, floods and droughts. The disasters in the state are occurring more frequently and penetrating into areas which were relatively free of disasters earlier. The prominent incidents which point to the fact that climate change is occurring in the state are as under:

- In 2001, twenty five out of thirty districts in the state were affected by floods. Districts in Western Orissa with no previous flood history were also affected. Floods have now become an annual affair for Orissa.
- In 2007, the coastal district of Balasore was affected by four floods virtually wiping out crops in the state and severely affecting human beings and livestock.
- The state suffered one of its worst droughts in 2001 which affected 11 million people in two thirds of the state's districts. The drought affected earlier drought free states such as Sundergarh and Kendrapara.
- In 1998, a heat wave in Orissa killed about 1500 people, mostly in coastal Orissa, an area known for moderate temperatures. The Titilagarh and Koraput belt comprising most of Southern and Western Orissa has experienced exceptional rise in daily maximum and minimum temperatures.
- In 1999, two cyclones hit the state in quick succession. The second cyclone lasted three days and affected 14 coastal distrcts. About 15 million people were affected and 50 per cent of the forest cover in Jagatsinghpur and Kendrapara was destroyed.
- Some districts in the state are affected by sea level rise. In Kendrapara district five villages out of a group of seven villages were submerged in the sea due to sea level rise. The two remaining villages, Kanhupur and Satbhaya, continue to face the danger of submergence as the sea continues to advance towards the villages.

Agricultural land in the coastal region would be vulnerable to inundation and salinisation. This would affect the paddy crop adversely. Orissa normally produces about 5 million tons of rice each year. Effect on agriculture would also throw a number of people dependent on it out of work.

A possible increase in cyclone intensity of 10-20 per cent against a rise in sea surface temperature of 2 degree centigrade to 4 degree centigrade is expected. It has been found that climate change has already intensified Asian monsoons and river flow.

In addition to Orissa, climate change impacts are also being noticed in the Sunderbans in the state of West Bengal. Although the area forms a small part of the state, it enjoys the status of being one of World Heritage sites due to its unique ecosystem. Sunderbans is one of the world's largest ecosystems spread over an area of 9,630 sq. km. It consists of 102 low-lying islands, 48 of which are inhabited. It is also the habitat of the Royal Bengal tiger. It has been estimated that 1 m rise in sea level will submerge 1000 sq. km of land in the Sunderbans. In the last 20 years four islands namely Bedford, Lochahara, Suparibhanga and Kabasgadi have submerged rendering 6000 families homeless. Scientists have estimated that one of the largest islands namely Sagar Island will lose 15 per cent of its habitat area by 2020. Communities are already responding by adjusting the timing of cropping, changing crop types, increasing rainwater harvesting and increasing island mud barrages.

West

Climate change impacts have also affected the urban and rural areas in Western India. The effects have been more pronounced on the Eastern coast as more climatological disturbance is seen over Bay of Bengal than over the Arabian Sea. However, the effects on Agriculture and on communities on the Western coast are visible.

- In the Western Indian state of Rajasthan pearl millet is the main crop. The production of pearl millet is reduced by 10-15 per cent due to a 2 degree rise in temperature.
- Rajasthan was affected by severe droughts in 2002. The drought affected 32 districts with a rainfall deficit of 53.4 per cent.

- The effects of climate change are also visible in Jhalawar, Rajasthan. The area has been experiencing droughts since the last four years. Agricultural yields have also reduced substantially, particularly for the majority of farmers who lack irrigation facilities. The medium farmers are coping by sale of available stock. The only option available to landless labourers is migration from the area. Small and marginal farmers cope in a number of ways which include sale of cattle, shift to other crops, labour and migration.
- Maharashtra too has been facing the effects of drought. In 2001, 20,000 villages in the state were affected by drought. The number of villages affected in 2002-03 and 2003-04 was 33 and 11 respectively.
- In 1994, the summer temperatures in Western India reached 50 degree centigrade providing favourable conditions for disease carrying vectors to breed. In the same year, there was a plague outbreak in Surat, Gujarat. In the period following summers of the same year Surat experienced heavy rains and was hit by a malaria epidemic. Hot and humid climate is ideal for disease carrying vectors.
- Coastal states of Maharashtra, Goa and Gujarat face a grave risk due to sea level rise. It has been estimated that a 1m rise in sea level will affect 7 per cent of the population in Goa and also affect 13 lakh people in Maharashtra. Besides actual inundation, a large number of people would be vulnerable to regular flooding also.
- In future, further negative impacts on agriculture are expected due to changes in soil, pests and weeds. In addition to above soil in the Western parts of India is prone to salinisation. This will further degrade the soil as the soil loses its moisture due to increase in temperature.

North

In the Northern part of India, impacts of climate change have been on water availability and agriculture. The average temperatures are also on the rise in these areas. The water scarcity is occurring due to receding glaciers in the Himalayas. The wheat yields and apple yields are also being affected due to rise in temperature. The rise in temperature raises issues for food security and livelihoods.

In Kullu valley, apple production in 1995 was one fourth of peak production year of 1988-89. In 1989-90, the production of apples reduced to 50 per cent of previous year's all-time-high production. The production has been declining since then. There have been seasonal shifts in the region due to climate change which have had an impact on the apple crop.

- In Kashmir global warming is affecting agriculture and water security. It has been found that temperature in Kashmir region has risen by 1.45 degrees centigrade and in Jammu by 2.32 degrees centigrade. IMD's monitoring reveals increase in temperature in Jammu and Kashmir valley with increase in maximum temperature of 0.05 degrees centigrade per year.
- The water level in almost all streams and rivers in Kashmir has decreased by 2/3rd during the last 40 years. Groundwater level has dropped by 1/3rd in the upper areas of the region.
- Snowfall has also undergone a change with December and January receiving scant rainfall and February and March witnessing heavy snowfall. While winter and spring water run-off has clearly increased due to early meltdown of snow.
- It has been seen in the valley that there has been a reduction in snowfall and increase in the frequency of cloudbursts.
- Agriculture in the valley has been adversely affected due to climate change.

Yields of paddy and maize have reduced due to increase in temperatures. The almond crop has also suffered damage. Due to reduced water in streams, maize and cherry yields have been reduced by half. Thus, it is seen that climate change has major implications for water security, food security and livelihoods in Kashmir.

South

In Southern India, Andhra Pradesh is an established climate change hotspot. The State has a long history of disasters. The coastal areas of the state are affected by cyclones and the areas in the interior are affected by droughts and heat waves. Certain districts namely, Ananthpur, Chittoor, Cuddapah, Hyderabad, Kurnool, Mehboobnagar, Nalgonda and Prakasam are chronically drought prone. In addition to droughts heat waves have become a major problem for the state since 1994. From 1986 to 1993, the heat waves were of a moderate nature and continued for a maximum period of seven days. Since 1994, the heat waves have become more severe and frequent and continue for a longer period.

- In 2003, 22 out of 23 districts in the state were affected by heat waves. In the same year 80 per cent of the state faced a drought.
- In 2003, drought in various parts of Andhra Pradesh forced the government to set up gruel centres in the state to help the villagers survive since even tea or tubers were not available. Such gruel centres were also set up in the districts of Ananthpur and Mehboobnagar. A large number of villagers migrated from these districts to nearby cities
- In 1997 and 1998, the spell of moderate to severe heat waves continued for 19 days.
- In May 2002, severe heat wave conditions were recorded in the Coastal districts of Andhra Pradesh. The heat waves occurred in two spells of six days and seven days. It has been seen that earlier heat waves occurred only in interior districts of Telangana. They have now shifted to the coastal areas.
- In 2003, the state experienced another heat wave with severe heat wave conditions lasting for a period of 27 days. The prominent coastal areas affected were Kavali in Nellore district and Ongole in Prakasam district.

It is seen from the above incidents that climate change is affecting areas which were free from such disasters and it is hitting these areas with unprecedented severity.

Increased temperatures would have an adverse impact on health of individuals due to increased heat stresses. It has been estimated by scientists at IIT, Delhi that temperatures over India have increased by 0.4 degrees centigrade over the last century. The monsoon rainfall is expected to increase by up to 30 per cent by the end of 21st century, as a result of climate change, which again would affect agriculture which is largely dependent on rains. According to World Bank, a 7 per cent increase in rainfall accompanied by 2 degree centigrade rise in temperature will reduce net agricultural revenues for the whole country by 12.3 per cent. The climate change models are expected to lead to a reduction in soil moisture leading to a reduction in growth of teak forests from 5.4 cubic meters per ha. to 5.07 cubic meters per ha. and productivity of deciduous forests from 1.8 cu. m per ha. to 1.5 cu. m per ha.

Table: Effect of 1-m Sea Level Rise on Coastal Area and Population

State/Union Territory	*Costal area (million hectares)*			*Population (million hectares)*		
	Total	*Inundated*	*Percentage*	*Total*	*Affected*	*Percentage*
Andhra Pradesh	27.504	0.055	0.19	66.36	0.617	0.93
Goa	0.37	0.016	4.34	1.17	0.085	7.25
Gujarat	19.602	0.181	0.92	41.17	0.441	1.07
Karnataka	19.179	0.029	0.15	44.81	0.25	0.56
Kerala	3.886	0.012	0.3	29.08	0.454	1.56
Maharashtra	30.771	0.041	0.13	78.75	1.376	1.75
Orissa	15.571	0.048	0.31	31.51	0.555	1.76
Tamil Nadu	13.006	0.067	0.52	55.64	1.621	2.91
West Bengal	8.875	0.122	1.38	67.98	1.6	2.35
		Union Territory				
Andaman and Nicobar Islands	0.825	0.006	0.72		0	0
India	**139.594**	**0.571**	**0.41**	**416.74**	**7.1**	**1.68**

Note: Coastal area and population are based on the 1981 and 1991 census.

Climate Change Induced Disasters

According to the IPCC report, the main threats to the urban populations and physical assets of developing cities, affected with more or less intensity based on the actual climate changes, which induced natural disasters, are the following:

- ***A Rise in Sea Level***: This is the most fundamental challenge that urban settlements face from global warming. The threat will likely increase due to the ongoing influx of people and economic assets into coastal zones. At risk are entire sections of coastal cities and their infrastructure, beaches subject to erosion, river floors in estuarine zones subject to sedimentation, and wetlands and tidal flats subject to flooding. Furthermore, groundwater risks increased salinization, and coastal aquifers risk diminishing, affecting fresh water supplies and peri-urban agriculture.

- ***Tropical Cyclones***: Increasingly frequent and intense tropical and extra-tropical cyclones will likely cause severe wind damage and storm surges which, compounded with a rise in sea level, are expected to become a severe problem for low-lying coastal regions and cities. Ports and other coastal infrastructure are especially at risk.
- ***Flooding and Landslides***: Expected increases in the scale, intensity and frequency of rainfall in most developing countries will severely strain or overwhelm the storm drainage systems of many urban centres. This could lead to periodic flooding of low-lying areas as well as landslides and mud-slips on geologically unstable slopes, often subject to informal settlements. Cities built next to rivers and on reclaimed lands in riverbed plains will be prone to additional inundations.
- ***Water Quality and Shortage***: Urban flooding damages water treatment works and flood wells, pit latrines and septic tanks. Sewage treatment systems and solid waste disposal areas can also be affected, contaminating water supplies. Where overall rainfall decreases, droughts will likely compromise the replenishment of the water tables, the normal sources of water supply. Doubling of CO_2 would lead to a global warming 1.5 to 4.5 °C causing droughts and forest fire. Water resources will be affected as precipitation and evaporation patterns change round the world.
- ***Heat and Cold Waves***: Intense episodes of thermal variability could severely strain urban systems by adding an environmental health risk for more vulnerable segments of the population, imposing extraordinary consumption of energy for heating and air conditioning where available, and disrupting ordinary urban activities.

Climate models predict that the global temperature will rise by about 13.5 °C by the year 2100. This phenomenon is known as global warming which may lead to climate change. This projected change is larger than any climate change experienced over the last 10,000 years. It is based on current emissions trends and assumes that no efforts are being made to limit green house gas emissions.

Natural hazards are an important cause for the loss of land in the coastal region of India. The deltaic regions of West Bengal, Orissa and Andhra Pradesh are especially vulnerable to cyclones and the storm surges associated with them. The recent super cyclone of Orissa provides an example of the type of damage that can happen. In particular, the interaction between a cyclone induced changes and the river discharges over the Hoogly estuary needs further research.

Indian Mitigation Policy on Climate Change

Climate change is having an impact on India as well. These changes may include decrease in crop productivity, water availability, increase in vector-borne diseases, extreme weather events like floods, cyclones and droughts and changes in perception patterns. In order to advance the preparedness of the country to deal with climate change, at present, the following activities are being undertaken:

- Capacity building at appropriate levels for taking Climate Change considerations in Social, Economic and Developmental Planning.
- Impact assessment including impacts on Food and Water Resources, Eco-system and Bio-diversity, Human Settlements and Human Health.
- Promotion of Scientific and Technological Research and Systematic Observation with a view to furthering understanding of Climate Change.
- Education, Training and Public Awareness.
- Enhancement of International Cooperation in pursuance of the objectives of the United Nations Framework Convention on Climate Change (UNFCCC).

India figures among the top ten contributors to greenhouse

gas emissions. However, the current per capita gross emissions in India are only one-sixth of the world's average. India is highly vulnerable to climate change as its economy is heavily reliant on climate-sensitive sectors like agriculture and forestry and its low-lying densely populated coastline is threatened by a potential rise in sea level. India ratified the Kyoto Protocol in 2002. Being a developing country, it was not required to cut emissions, but it has developed certain mechanisms to reduce emissions. The primary legislation in India is as follows:

- National Hydro Policy, 1998—Outlined strategies for hydro power potential.
- Energy Conservation Act, 2001—Established Bureau of Energy Efficiency.
- A National Biodiesel Policy is also being considered to encourage use of Biodiesel.

Clean Development Technologies

An attempt is being made all-over the world to adopt technologies which would help to reduce energy needs and are non-polluting. The focus, at present, is on the following technologies:

- Wind energy.
- Solar power.
- Biodiesel.
- Hydro power.

India also established CDM-India board in 2003 and a large number of CDM projects have been set up in India.

Role of Different Sectors in Mitigatory Measures of Climate Change

High Powered Committee (HPC) adopted a strong process oriented and participatory approach involving different sectors of the society like NGOs, Media, Police and Security Forces, Communications and Governments including Administrative Training Institutes (ATIs) at all levels, etc. as against an academic and technical approach (HPC Report, 2001).

The Central Government especially National Institute of Disaster Management (NIDM), Ministry of Home Affairs and all ATIs of State Governments through their Natural Disaster Management (NDM) faculties may start the training programmes on climate change by incorporating the contribution of different research centres working in the country and abroad. They may also conduct the training programmes on environmental Laws for the officers of Central Pollution Control Board (CPCB), State Pollution Control Board (SPCB) and Public Prosecutors of States and UTs as they may be helpful in exploring the knowledge of various Acts of Environmental Protection, Water and Air Pollutions, Wildlife, Fisheries and Forests, etc. among the people and force them to follow these Acts strictly.

The role of the Electronic Media has during recent times emerged as a major component in the field of global climate change. At the same time, the role of the Print Media cannot be taken on a lower level of importance because it continues to be the medium of mass media in many parts of the Indian society still un-reached by the electronic media. The Educational Bodies apart from School to University level have been implementing educational and awareness programmes on environment including climate change.

Climate Change Adaptation

Various NGOs are working to help communities to adapt the modified climatic conditions. One admirable effort in this direction is by a NGO, DSCRC, which has introduced an innovative rainwater harvesting system in Purulia district of West Bengal. This system has helped farmers in the area to cultivate an additional winter crop. Another effort to address a different aspect of the problem has been made by SEEDS India which constructed 300 houses for the villagers after the Floods in Barmer, Rajasthan. The traditional design of the houses was retained with the difference that the new houses were made out of mud stabilized with 5 per cent cement and had proper foundations and structural members. Services of local masons were utilized to train them in the new construction techniques. In addition to above, NGOs have also been working with communities on other aspects such as use of renewable energy to reduce energy and crop diversification but the important issue is that climate change adaptation calls for a change in traditional lifestyles of individuals or communities. The reality of climate change is before us. The choice is between adaptation and annihilation.

India: Climate Action Plan

The then Indian Prime Minister, Dr. Man Mohan Singh released the National Climate Action Plan on 30 June, 2008, with eight missions, the emphasis on solar energy is bound to generate the most excitement as this sector has been lagging behind other segments of renewable power sources. The eight missions are:

1.	Solar Energy	1,000 MW solar power by 12th Plan end
2.	Energy Efficiency	10,000 MW saving by 2012
3.	Energy Efficiency target for industries	Steel, Power, Textile, Fertilizers, Cement, Chlor-alkali, Aluminium, Paper and Pulp, Railways, etc.
4.	Minimum target of 5 per cent renewable energy for power grids	
5.	Nuclear power (Clean Energy)	part of mitigation measure
6.	Critical data/information sharing	
7.	Recycling from automobiles	
8.	Coal thermal power plants to be closed	5,000 MW Coal thermal plans will shut by 11th Plan. Additional 10,000 MW by 12th Plan.

Conclusion

The global heating process is already engaged and we cannot stop it in the near future. Indeed, thermal inertia is protecting us now by retarding the effects of trace gas build-up, but this same inertia will insure continued perturbations for centuries, even after correction becomes efficient. What is most important now is synthesis, the overall appreciation of the problem of global climate change and possibilities for correction, bearing in mind regional constraints and desirable changes. This should be the concern of the nascent international community of person interested in the survival and well-being of the biosphere. Undoubtedly, the CDM created under Kyoto Protocol (1977) to the UNFCCC, is important for India as it allows developed countries to not only use CER credits from project activities in developing countries but also assists developing countries in achieving their sustainable development objectives.

The Eighth session of the Conference of Parties (COP-8) to the UN Convention on Climate Change (October 23 to November 01, 2002), New Delhi, ended here with a Delhi Declaration which failed to break new ground to stabilize the world's climate but firmly projected the view of the developing countries. However, this meet successfully resolved the technical parameters necessary for the Implementation of the Kyoto Protocol (1997) on Green House Gases reductions. The Delhi Declaration gave primacy for the implementation of the clean development mechanism in the climate change process. Hence, developing countries refused to take on any commitments for reductions in GHG emissions. In the implementation of the National Action Plan on Climate Change, there is need of coordinating efforts of all concerned ministries. As the Climate Change issues are related to the environmental management and better resource planning, good governance can lead towards sustainable development and reduce the impact of climate change. There should be legal support to the administration for climate change adaptation.

14

Green Computing

Best Practices to Save Environment

S. Arumuga Perumal

Introduction

Green Computing is a study and practice of designing, manufacturing, using and disposing of computers, servers and associated subsystems such as monitors, printers, storage devices, and networking and communication systems effectively and efficiently with minimum impact on the environment. Modern Information Technology systems rely upon a complicated mix of people, networks and hardware; as such, a green computing initiative must cover all of these areas as well.

Pathways to Green Computing

To comprehensively and effectively address the environmental impacts of computing/IT, we must adopt a holistic approach and make the entire IT lifecycle greener by addressing environmental sustainability along with the following four complementary paths:

- **Green Use** reduces the energy consumption of computers and other information systems as well as using them in an environmentally sound manner.
- **Green Disposal** reuses old computers and properly recycling unwanted computers and other electronic equipment.
- **Green Design,** designs energy-efficient and

environmentally sound components, computers, servers, cooling equipment, and data centres.

- **Green Manufacturing** manufactures electronic components, computers, and other associated subsystems with minimal impact on the environment.

Energy Saving

The development of Information and Communication Technology has led to a mounting pressure for big internet companies to reduce their energy usage. The companies face pressure from government and others concerned with the environment to reduce their carbon footprints. As computers have become more numerous and the cost of hardware has declined relative to the cost of energy, the energy efficiency and environmental impact of computing systems and programmes has received increased attention. The algorithm developed by the researchers does not necessarily reduce the use of energy. It merely helps companies better manage their energy costs. The efficiency of algorithms has an impact on the amount of computer resources required for any given computing function and there are many efficiency trade-offs in writing programmes.

A study by Alex Wissner-Gross, a physicist at Harvard, estimated that the average Google search released 7 grams of carbon dioxide (CO_2). However, Google disputes this figure, arguing instead that a typical search produces only 0.2 grams of CO_2. The Google data centre in Belgium relies on ambient cooling to keep its facilities under a certain temperature instead of firing up expensive air-conditioning systems. On days when the weather is too warm, the servers in the data centre simply shut down. The researcher working on the project said that the energy-routing algorithm is an extension of this idea.

So an urgent solution may be needed to address end-user satisfaction, management restructuring, regulatory compliance, and return on investment (ROI). There are also considerable fiscal motivations for companies to take control of their own power consumption: "of the power management tools available, one of the most powerful may still be simple, plain, common sense."

The literature review from the Advanced Configuration and Power Interface (ACPI), an open industry standard, allows an operating system to directly control the power-saving aspects of its underlying hardware. This allows a system to automatically turn off components such as monitors and hard drives after set periods of inactivity. In addition, a system may hibernate, where most components (including the CPU and the system RAM) are turned off. ACPI is a successor to an earlier Intel-Microsoft standard called Advanced Power Management, which allows a computer's BIOS to control power management functions.

Some programmes developed to save the energy allow the user to manually adjust the voltages supplied to the CPU, which reduces both the amount of heat produced and electricity consumed. This process is called undervolting. Some CPUs can automatically undervolt the processor depending on the workload; this technology is called "SpeedStep" on Intel processors, "PowerNow!/"Cool'n'Quiet" on AMD chips, and LongHaul on VIA CPUs.

Managing energy and its costs is likely to be big business in the coming years for technology companies. Other high-tech companies, including IBM, Intel, and several start-ups, are also ramping up to develop sensors and other technology that alert power companies and their customers about energy usage in an effort to balance usage and reduces costs.

Battery free operation is a key step in paving the way for truly Green Wireless systems that do not adversely impact the environment. Battery powered wireless sensor networks can be installed at reduced cost. Wireless Green powered sensor network requires battery free wireless solution. It can be achieved by creating Ultra Low Power (ULP) WSN node that can utilize ambient environmental energy sources such as light, thermal, motion and vibration. Energy harvesting is the process by which energy is readily available from the environment. To implement this technology we need software to monitor energy harvesters and optimizing the use of harvested energy which enable sensors to operate reliably in a battery free environment.

Parameters in Production of ICT• Products

Designing, manufacturing, using and disposing of ICT products along with associated subsystems efficiently and effectively with minimal or no impact on the environment can be termed as Green ICT. As per our analysis, the parameters proposed for Green ICT are Energy efficiency, Carbon emissivity, Internal Heat Generation, Radiation of EMF, Scalability and Biodegradability.

E-waste

When electronic devices are judged as no longer useful and are discarded, they become electronic waste or e-waste. In todays electronic world, e-waste is the biggest and fastest growing manufacturing waste. E-waste includes computers and peripherals, cell phones, PDAs, digital cameras, etc. As per the survey, Waste Electrical and Electronics Equipment (WEEE) account for 90 per cent of the generation with details:

- Household appliances—42.1 per cent,
- ICT products—33.9 per cent
- Consumer electronics—13.7 per cent.

In India, rough estimate suggest more than 1,45,000 tons annually. So we have to raise the awareness of hazards of e-waste and stimulate the necessary debate by providing information about the sources/causes of the problem, effect of e-waste substances on human health and propose "E-waste take back system" and thereby turning green.

Recycling and Environment Friendly Design Solutions

The PC manufacturers need to use more sustainable and less toxic recycled plastics and bio-plastics in their tower and monitor casings. They ought to use a biodegradable corn-based plastic case in its printers and use wood and bamboo wood encased desktops in computer towers, monitors and mouse controls. There is an urgent need to eliminate CRT monitors that emit high toxic radiations.

The LCD screens are also not completely free of toxic materials. The energy efficient and toxic free OLED technology would become cost effective. OLED replaces LCD screens as industry standards. The e-ink and IMOD are the two other monitor related technological breakthroughs that has the potential to have a significant impact in future green technology. The packaging solutions to pack printers in recycled plastics and recycled content paper board. Ink and toner cartridges in reusable air cushioning rather than molded end caps should be preferable.

Health Hazard on Youngsters

In today's electronic environment, it is very difficult to find people who are not using TV, cell phones, computers, etc. But there has been an increase in the complaints of headache, fatigue, sore eyes, hot ear, pregnancy problem in women, disappearance of certain species of birds in Metropolitan cities and in extreme cases even results in development of malignant cells in the upper part of the body which results in cancer. There associated impact is worse on the health of both human beings and other species of animals and birds with the use of these technological equipment. Radio frequency sources become integrated part of today's lifestyle. Our exposure to electro pollution, referred as EMFs has increased dramatically. The EMF energy is observed by the brain cells lead to thermal loads and temperature gradients. The continuous exposure to these radiations can cause DNA damage. The damaging effects of the radiations can produce a wide range of physical symptoms. Some studies suggest that potential health hazards could be linked to excessive exposure to high power densities of non-ionizing radiations. These health hazards include cancer, tumors, headaches, fatigue, etc.

Consumer Awareness

The development of Information Technology leads to social engineering attacks which occur either electronically or in person. The danger of social engineering is that it invalidates the protection provided by corporate security products and makes your organization vulnerable to attacks and serious losses. Social engineering succeeds by finding the weakest link in the company. Therefore, an important part of prevention is to report any suspicious conversation or solicitation by an outsider. If people suspect they are being socially engineered and they immediately report this fact, then perhaps the social engineer can be defeated. So we are in a position to keep the company workers/customers knowledge base up-to-date on any new dangers that they should be cautious about.

The different security measures that are to be considered in financial sectors are to share the fundamental concept behind the security technology, to share the new innovative concept and to understand the relative advantages and limitations of different approaches. There are many technologies available to counteract intrusion, but currently no method is absolutely secured. The most dangerous frauds that causes in day to day banking activity is phishing, a criminal activity using social engineering techniques. The protection afforded to banking information is usually expressed in terms of confidentiality, integrality, availability and non-repudiation. Authentication has a significant contribution to provide all these services. People often think of computer security in terms of security products. But the main element of security is clearly the human element.

Dispose Digital Media Safely

There is a dire need to think of ways and means to reduce the environmental impact of replacement of the old computer and safeguard your private information that is stored in storage devices. One of the most forgiving features of computers is the ability to undelete a file. Even though a file has been deleted, it is possible to unerase it. Recovering a deleted file is possible since the system does not physically erase the file; it simply marks the storage location as available. Not only this, even if you reformat a drive, much of the data still can be recovered. If you simply throw a computer away, you seriously compromise of all your confidential files, trade secrets and proprietary data. If the discarded computer falls into the hands of a person who knows how to access the hard drive, then he can freely retrieve the confidential data and use it without a trace, such as your name, birth date, social security number or credit card numbers. To reduce the environmental impact of our old computer we can donate it to a recycling centre instead of throwing it in the trash. The best thing to do with the hard disk drive is physically destroy it.

If we want to make sure third parties are not able to recover deleted data, the best way to do this is to degauss the media, Degaussing is done using a special hardware and subjects the media to very strong electromagnetic fields and permanently erases any data found on the storage medium. Degaussing does not work on WORM or CDROMs. Another method used for this is data overwriting or zeroization. This involves successive writings of random data on the drive, making the data unrecoverable. This overwriting is done using special softwares.

The most effective method is the actual destruction of the storage media. It has to be kept in mind that for the destruction to be truly effective, the media has to be completely destroyed. It is not enough to simply hit a hard drive a few times with a hammer or drop it on the floor.

The following are some of the ways to make the computers more environment friendly. Modern computers are power hungry things. There are many things we can do to reduce the amount of power they use. As a general rule, laptops use less than Desktops, and LCD (Flat screens) Screens use less than CRT (big fat screens).

Tips for Green Computing

- Switch off the computer when it is not in use.
- If you are going to be away from it for a short time then turn off the monitor. One can also adjust the power management using the control panel to turn off the hard drive and put the system into standby after a set amount of time.
- For more advanced users, remove unnecessary fans and drives. If you have a spare CD-ROM, used once in a blue moon, unplug it. Also if the computer is only used for office work or checking e-mails, unplug excess devices not in use.
- Instal third-party power management programmes like Data Synergy PowerMAN or Snap Verdiem Surveyor. These programmes allow power management to be centrally configured which can be very important in larger organizations.
- Print wisely—Only print what you really need to have a paper copy and resist the urge to print 'all pages' when you only need what is on one particular page. There is software available that can help with this, or simply use the "print preview" to see which pages you really need and only print those pages.
- Refill Ink-jet cartridges and laser toner; it is cheaper and does not add to landfill.

- Switch off Cable Modems, Routers and Wifi Access Points when not in use/overnight.
- Replace the internal hard drives with a low energy solution such as Western Digital's Green Drives.
- An external hard drive is also useful to backup your content, or to use as simple add-on storage capacity. Like the SimpleTech redrive, look for one that is energy efficient and Energy Star qualified, is made from sustainable materials such as bamboo and aluminium and is conscious about its packaging.
- Use a Google alternative, such as Smallest Google, saving energy, time, and bandwidth.

Tips for buying a New PC

- Buy a Laptop. Laptops use less power than a desktop.
- Buy a computer according to your needs. If you are only using it for simple tasks do not buy the most powerful available.
- Choose the latest Intel or VIA processors over AMD as they use less power. VIA is the best for the environment although not as powerful.
- If you are going to build your own PC pick the parts carefully. Instead of getting two 1gb chips, buy one 2gb chip. Build the computer to your needs do not install unnecessary parts if you will never use them. Also consider certified 80 plus energy efficient power supplies 80 PLUS Programme.
- If you are concerned about hazards chemicals used in pc components, consult Greenpeace's Green Electronics Ranking or The Restriction of Hazardous Substances (RoHS) in electrical and electronic equipment.
- Choose Corded Keyboard and Mouse over wireless.
- As a general rule, ink-jet printers use much less energy than laser ones however the inks in the inkjet printers spoil clean paper recycling processes.

Green IT Tips

- Employ a policy of only printing when it is essential.

- Go paperless with incoming faxes by having them sent to you through e-mail, you can save many thousands sheets of paper each year. What makes this even more efficient is that the person who receives them can easily delete spam and quickly forward faxes to the relevant person in your office.
- Print double sided when possible. If you have employed a double sided printing policy you would instantly save half your paper consumption and reduce your printer paper costs by 50 per cent.
- Use recycled paper for printers and faxes.
- Use Blackle rather than Google. It will give you the same search results but the black screen uses less energy than Google's white background.
- Use an energy efficient server.
- Avoid buying IT products with excessive packaging or packaging that is non-recyclable.
- Calculate and offset your carbon footprint. http://www.carbonfund.org/

Conclusion

Due to the fast increase in the growth of population and increasing quantities of electrical and electronic products, there are serious issues that need to be addressed, not just at the end of the life but throughout the product life cycle. The key challenge is to develop further the procedure, processes and materials that will enable efficient handling and recycling of e-waste. The Green ICT can be acquired by encouraging the production of ICT products with designing, manufacturing, using and disposal of ICT products with minimal or no impact on the environment. Creating awareness among the youngsters about the EM radiations, methods to prevent the EM radiation effect and probable methods of reducing these health hazards will make environment green in the near future.

REFERENCES

Green Computing, wikipedia, the free encyclopedia
Inter PC energy-Efficiency trends and technologies

San Murugesan, (2007), "Going Green with IT: Your Responsibility Toward Environmental Sustainability." Cutter Consortium Business-IT Strategies Executive Report, Vol. 10, No. 8, August.

San Murugesan (2008), "Harnessing Green IT: Principles and Practices", IEEE *IT Professional,* January-February, pp. 24-33.

www.greentechnology.com

www.wikipedia.org/wiki/electronic_waste

15

Alternate Building Materials

An Effective Method to Curtail Carbon Emissions in Indian Scenario

Sabu. M. Simon and *Diraviaraj Maria Susai*

Introduction

The biomass burning and its implications on atmosphere, climate and biosphere are widespread and extensive. The Third World countries are considered to be the main contributors of biomass burning in the international scenario. The magnitude of emissions from biomass burning in rural settings is not well estimated as it is carried out by women, farmers and livestock rearers and as such there is no estimation of amounts of biomass burnt by them.

Biomass burning also contributes significantly to the increasing air pollution in India. Calculations of regional emissions from biomass burning show that a cumulative emission of 0.12Tg of SO_2, 0.71Tg of NO_2, 270Tg of CO_2, and 17Tg of CO is being emitted from South Asia, of which more than 70 per cent is contributed by India (Streets *et al.*, 2003). In agriculture and economy, biomass burning serve a variety of purposes including (i) cleaning of forests and brush land for agricultural use, (ii) Control of brush, weeds and litter accumulation on grazing and crop lands, (iii) nutrient regeneration in grazing and crop lands, (iv) control of fur accumulation in forests, (v) production of charcoal for industrial and domestic use, and (vi) energy production for cooking and heating (Meinrat O. Andreae, 1991). Apart from the purposes mentioned above, the energy requirement in brick making industry to fire the brick kilns also causes huge quantity of biomass burning.

Fuel wood is the major source of energy in developing countries. About 50 per cent of the fuel wood is used for cooking, 30 per cent for domestic heating and the remaining 20 per cent for various other purposes (Meinrat O. Andreae, 1991). Buildings are responsible for more than 40 per cent of the global energy use and one third of the global greenhouse gas emissions, both in developed and developing countries (UNEP-SBCI, 2009). India is one of the fastest growing construction sectors in the world.

Brick Making Process

Locally available clay sand is used for the brick making process. Clay bricks require a soil with clay and sand combination of not less than 50 per cent by weight of the former and fired at high temperature (800-1000 °C) to make it dense and hard (Kumar and Visvanathan, 2003). The brick making process involves sand preparation, mixing, molding, drying and firing.

Brick Kilns in India and their Performance

Different type of brick making technologies like Clamps, BTK, Hoffman, Chamber, Tunnel, VSBK etc., are being employed in India for the brick making process. The energy consumed and the pollution created differ from kiln to kiln, depending upon the technology used in the kilns. The energy consumed for firing 1 kg of fired brick is the better way of comparing the performance of the brick kilns. In most of the Indian kilns, the percentage of heat that is wasted generally varies from 40-90 per cent, the reason for which is the loss in flue gases and convection and radiation heat loss from the kiln surface. (Sameer Maithel *et al.*, 2009) (*see* Table 15.1).

Table 15.1: Approximate Energy Consumption of Different Brick Making Kilns

S. No.	*Technology*	*Specific Energy Consumption (MJ/khg of the product)*
1.	Intermediate Kiln	3-11
2.	Scotch Kiln	1.5-7
3.	Bull's Trench Kiln	1.8-4.2
4.	Hoffman Kiln	1.5-4.3
5.	Tunnel Kiln	1.5-2

Source: PSG, 1998.

Study Area

The study area selected was Dindigul block of Dindigul District of Tamil Nadu. An area of 12 kilometers of radius of Dindigul block, with Vedasandur on the northern side, Nilakkottai on the southern side, Reddiyarchathiram and Attur on the western side and Shanarpatti and Vadamadurai on the eastern side.

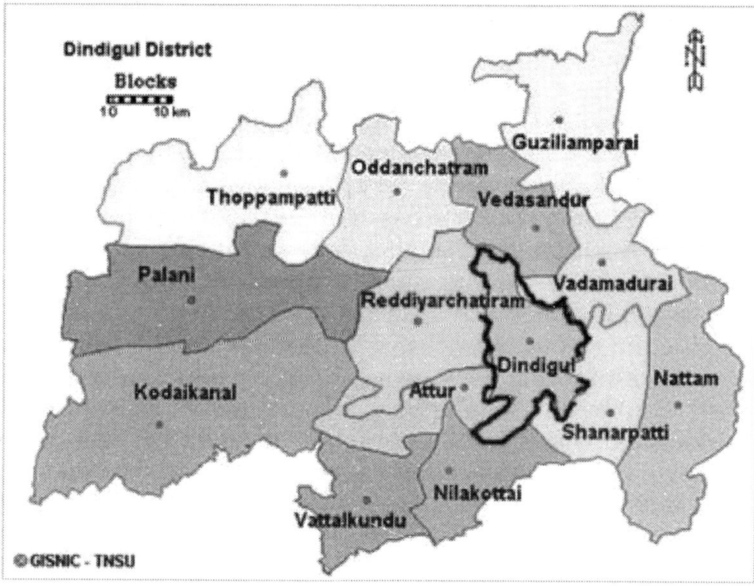

Soil Types and Usage

The major soil types in the district are 1. Red Soil, 2. Red Sandy Soil, and 3. Black cotton soil. Red soils are prevalent in Palani, Nattam and Ottanchathiram, while Red sandy soils are prevalent in Nilakkottai, Dindigul and Vedasandur. Black soil is found in all Taluks except Kodaikanal. The major land area of 2,53,505 ha is being used for Agricultural purposes, 1,38,923 ha remains as forest, 65,184 ha of land is put to non-agricultural uses and 36,210 ha of he land is barren and uncultivable. The remaining land area of ,32,842 ha of the total 6,26,664 ha is grazing land, fallows, ultivable waste etc. (District Ground Water Brochure, 2008). The ndom pH analysis in the study area show a pH, which is neutral to slightly acidic in the range of 6.9 to 7.8. A moisture content of 20.5 to 23.1 was also recorded in the random sampling in the Dindigul Municipality (Municipality Public disclosure data).

Brick Kilns in the Study Area

There are 89 medium and small size brick kilns operating in the study area of 12 km radius. Among this, 19 are chamber brick producing kilns, which produce superior quality bricks and the remaining 70 are country brick producing kilns. Consumption of the country bricks is high in the area as in other parts of the country. All the brick kilns have the maximum production capacity of 1,50,000 bricks per cycle and the kilns would be producing 50,000 to 1,50,000 bricks per cycle based on the season and demand. In all the kilns, traditional chamber systems with movable chimnies are being used. Depending on the number of bricks burnt, it takes one month to three months for the completion of one rotation of production of cured bricks. Based on the time taken for the completion of curing, it requires 100 to 300 man-days of manpower utilization for the completion of the process. At an average, the bricks cured would be sold out within a maximum time span of 21 days. All the brick kilns together produce an average of 5,34,00,000 bricks in a year. For the purpose, an average of 21,707 man-days is utlilised.

Biomass used for Brick Firing

Non-woody biomass of *Prosopis juliflora* is widely used in the brick kilns in the study area. *Prosopis juliflora* is a shrub or small tree and used as forage, wood and for environmental management. The tree grows to a height of up to 12 metres (39 ft) and has a trunk with a diameter of up to 1.2 metres (3.9 ft). Eack kiln requires 10 tonnes to 15 tonnes of raw wood for curing the bricks per cycle. Each brick would require approximately 2.5 kgs of raw firewood for the curing process. The 89 brick kilns in the area consume an average of 1068 tonnes of dry biomass for a cycle and the total biomass consumption in an year is 6,408 tonnes.

Manpower Utlilization in the Brick Kilns

The brick production is a labour intensive industry. In the 89 brick kilns in and around Dindigul, a total of 679 workers are employed. Around 45 per cent of the workers are migrants from other districts and the remaining are natives of Dindigul. It was observed that there is a lean period of 100 to 180 days in the rainy season and in the remaining period of the year, they have continuous work in the kilns and both men and women are employed in the kilns in Dindigul.

Carbon Content in the Biomass Burning

To estimate the carbon content of the biomass, the units of Carbon was derived from the quantity of biomass (tdm). A default value of 0.49 t C/tdm was applied for shrubby biomass and the carbon content of the biomass was measured as tonnes of carbon per tonne of dry matter.

GHG Emissions from the Brick Kilns in the Study Area

For the calculations of the GHG emissions, the methods narrated by Syed Ashraful Alam (2006) and default values in the A/R Methodological tool of UNFCCC/CCNUCC (2009) were used.

CO_2 Emission from Fuel Wood Burning

The total CO_2 emission from the brick kilns in and around Dindigul during the brick making process was calculated using the following formula:

Total fuel wood burnt (t dm) x Fraction of Biomass oxidized (0.9) × Biomass Carbon Content (0.49 t C/t dm) × 44/12 = Total CO_2 released from the fuel wood burning. To convert total C released into total CO_2 released, the ratio of 44t CO_2/12 t C was applied.

It was estimated that 6408 tonnes of biomass is burnt every year in the study area for making 5,34,00,000 bricks from the kilns. Applying the formula, it was calculated that 10,361.5 t of CO_2 is emitted every year from the 89 kilns for brick making.

Other GHG Emission from the Brick Kilns

The quantum of non CO_2 GHGs including CO, CH_4, N_2O as a result of biomass burning from the brick kilns was also calculated.

CH_4 emissions (tCH_4) = Carbon released (tC) x emission ratio (0.012) × 16/12. Applying the formula, it was observed that 45.75 t of CH_4 is emitted from the studied kilns in an year.

CO emission (t CO) = Carbon released (t C) × Emission ratio (0.060) × 28/12. It was observed that 395.6 t of CO is emitted in a year.

N_2O emissions (t N_2O) = Carbon released (t C) × Emission ratio (0.007) × N/C ratio (0.01) × 44/28. A total emission of 0.31 t of N_2O was also observed to be released from the brick kilns in the study area in a year.

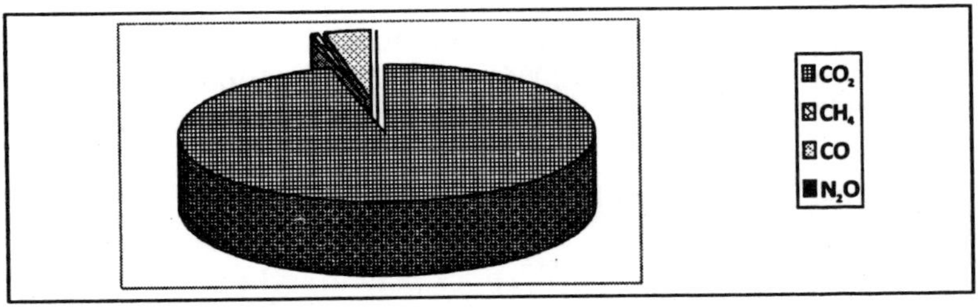

Fig. 15.1: Different GHGs emission from brick kilns.

It was observed that considerable Greenhouse gases are being emitted from the existing 89 brick kilns situated in a radius of 12 kilometers around Dindigul block. These brick kilns thus pose a threat to the climate and to the human health also.

Stabilized Mud Block Technology

In many of the rural houses, still raw earthen soil is used for the construction of walls but the usage is limited because of its low durability due to water penetration, erosion of wall just above the ground surface due to splashing of rain water and attack by termites. These limitations are overcome by Compressed Mud Block Technology, in which, the block density of the soil used is increased multifolds through compaction of mud mixed with a small quantum of cement or lime in a mechanical press. Stabilized Mud Block Technology is a simple, cost-effective, environment friendly technology developed by Centre for Science and Technology, Indian Institute of Science, Bangalore. It utilizes local materials and reduced energy consumption and thus, reduces the cost. The stabilization is always done on a dry weight basis. For the stabilization of the mud, cement is used at an average of 5 per cent for sandy soils or lime is used at an average of 6 per cent for clayey soils. The process of production involves sieving the soil, mixing the soil with sand and stabilizer such as cement and lime, mixing of optimum quantity of water, pressing the wet mixture into a dense solid block using a simple manually operated machine and curing the block for a period of 3-4 weeks by sprinkling moisture regularly.

Comparison of Stabilized Mud Blocks with Bricks

Compressive Strength

The compressive strength of stabilized mud blocks made from soils taken from different sources were tested at RVS College of Engineering and Technology, Dindigul, Tamil Nadu and Gandhigram Kattida Maiyam of Gandhigram Rural University. The sample mud blocks with a size of 23 × 19 × 10.5 cm were tested after 21 days of curing and showed an average compressive strength of 86.80 kg/cm^2.

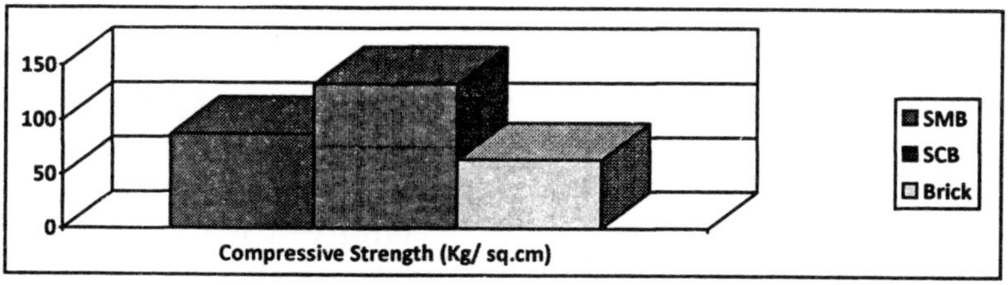

Fig. 15.2: Comparison of Compressive Strength of Stabilised Mud Block, Stabilised Crusher Block and Brick.

Simultaneously, blocks were also made of crusher stone dust, stabilized with cement. The stabilized blocks of crusher stone dust 30 × 14.5 × 10.5 size had shown a greater compressive strength of 133 to 210 kg/cm^2.

The country bricks from the local brick kilns were also subjected to compressive strength test. The average compressive strength was 63.4 kg/cm^2, which was observed to be significantly lesser than the Stabilised Mud Blocks (SMB) and Stabilised Crusher Blocks. So compressed mud blocks have sufficient load bearing capacity and no question of strength arises when used for single storied buildings in the rural settings.

Soil Degradation due to Brick making

Huge quantity of clayey soil is required for the production of bricks. This results in the loss of fertile top soil and soil erosion. Usually the soil is taken from fertile lands, which can be effectively used for agricultural purposes. The study area calculations show, it was calculated that an average of 3600 kgs of top soil is taken by the brick kilns for the making of 1200 bricks. So, in an year, for the making of 5,34,00,000 bricks from the area, a minimum of 1,60,200 tonnes of top soil is excavated. The excavated area is left barren and in the next rainy season, it becomes the breeding ground for the mosquitoes as well.

Whereas, the stabilized mud blocks can be made from any type of soil. The soil excavated for the foundation pits of the buildings can be effectively recycled into stabilized mud blocks. The de-silted pond soils or the left away soils during any excavation process can be recycled and reused into SMB and this could reduce the soil piles in any excavation process.

Conclusion

Over the years with an increase in building construction manyfold activities there is an increase in consumption of burnt bricks also. During preparation, baking and use of the bricks, considerable quantum of carbon is emitted. The compressive strength and durability of the Stabilised Mud Blocks were assessed and found to be an ideal alternative for the burnt bricks atleast for the single storied buildings. But a sample survey taken among the people, who are constructing houses in the area recently had revealed that majority of the common people are not aware of the SMB, which can be used for construction instead of burnt bricks. Even the people, who are aware of the technology, are doubtful about the strength and durability of the Stabilised Mud Blocks. Concerted efforts are required from the Government and other agencies interested to promote the eco-friendly technology to curtail carbon emission and also to minimize the construction costs.

REFERENCES

Central Ground Water Board, at http://cgwb.gov.in/District_Profile/Tamilnadu/Dindigul.pdf

Dindigul Municipality Public disclosure data at http://municipality.tn.gov.in/dindigul/dgl per cent20part-8_0.pdf. (Accessed on 29 September 2010).

District Ground Water Brochure, 2008.

Kumar, S. and Visvanathan, C, (2003). *Brick and Ceramic Sectors in Small and Medium scale Industries in Asia: Energy and Environment*, pp. 11-25.

Meinrat O. Andreae, (1991). *Biomass burning: Its History, use and distribution and its impact on environmental quality and global climate* in Joel S. Levine (ed). Global Biomass Burning: Atmospheric, Climatic and Biospheric implications: Massachusetts Institute of Technology, pp. 3-22.

PSG College of Technology, (1998). *Status report of the brick industry in India*. SMIs in Asia: Energy, Environment and Climate interrelations. India.

Sameer Maithel, Uma, R., Anil Kumar and Vasudevan N., (2009). *Energy Conservation and Pollution control in Brick kilns* at http://www.cosmile.org/papers/brick_energyconservationpollution control.pdf

Streets, D.G., Yarber, K.F., Woo, J.H. and Carmichael, G.R, (2003). *Biomass burning in Asia: annual and seasonal estimates and atmospheric emissions* at http://www.cgrer.uiowa.edu/EMISSION_DATA/publications/Streets_GBC_paper_revised_20030523.pdf.

Syed Ashraful Alam. (2006). Use of Biomass fuels in the brick making industries of Sudan: Implications for deforestation and greenhouse gas emission. Dissertation submitted to Department of Forest Ecology, Viikki Tropical Resource Institute, University of Helsinki.

UNEP-SBCI. (2009). *Buildings and Climate Change—Summary for Decision makers* at http://www.unep.org/sbci/pdfs/SBCI-BCCSummary.pdf. (Accessed on 11 September 2010).

UNFCCC/CCNUCC. (2009). *Estimation of the increase in GHG emission attributable to displacement of pre project agricultural activities in A/R CDM project activity* at http://cdm.unfccc.int/methodologies/ARmethodologies/tools/ar-am-tool-15-v1.pdf. Accessed on 14 September, (2010).

16

Sustainable Agricultural Production and Consumption Based on Total Quality Management

E.K. Satheesh

Introduction

Rising food prices is one of the greatest concerns of societies all over the world. This situation is considered to be a setback to the poverty alleviation programmes and overall health improvement scenario of the people living in underdeveloped countries. India is the second largest food producer in the world after China. Indian food market is estimated to be at US $ 182 billion and the retail food sector in the year 2008 was US$ 70 billion. This is expected to grow at US $150 billion by 2025.

The Union Government has already taken steps to prevent food shortage caused by the ongoing monsoon failure. At the time of independence, the contribution of agriculture to GDP was 70 per cent and employing nearly 85 per cent of the total population. Over a period of six decades, the contribution of agriculture to GDP was drastically reduced to 18 per cent. But the percentage of employment in agriculture more or less remains the same (70 %). India expects a GDP growth of 8 per cent per annum. As per Table 16.1, the average growth of GDP for the last three years is only less than 8 per cent. This indicates that the agricultural sector should grow at the rate of 4 per cent, but as per the Table 16.2, it is growing at less than 2 per cent. "We thus face many challenges on farm front. We have to run twice as fast to stay where we are. The current obsession with bricks in institution building should give way to nurturing brains. We will then have a bright agricultural future."[1] It indicates the need for capacity building in agricultural production. There should be proper forward integration aiming at the value addition of agricultural sector. In the given circumstances demand for increased agricultural production and quality food production is becoming the biggest challenge for policy-makers. The food processing sector should be given further impetus so as to make the agro-processing industry more sustainable and quality oriented.

Table 16.1: Economic Indicators (Growth %)

	2007-08	2008-09	2009-10
Overall GDP	9.2	6.7	7.4
GDP from Agriculture and Allied Activities	4.7	1.6	0.2
Foodgrain production	6.2	1.6	(-)6.9

Source: CSO GoI.

Table 16.2: Sectoral Growth Rates of GDP

	2005-06	2006-07	2007-08	2008-09	2009-10
Agriculture and Allied Activities	5.2(18.2)	3.7(17.1)	4.7(16.4)	1.6(15.6)	-2(14.6)

(Figures in parentheses indicate percentage share to total GDP.)

Source: CSO GoI.

Sustainable Agricultural Production in India

Agricultural production in India is increasing in absolute terms but its growth depends on monsoon availability and increased application of agricultural production techniques. Its growth is not sustainable and includes shrinkage of the fertility and the area available. "Farmers in many parts of the country are still operating under unsecured and unrecorded tenancy arrangements and declining farm size as a result of land fragmentation and proliferation of concealed tenancies will continue to put downward pressure on land productivity.[2] Agricultural production in India is not sustainable due to the ever-increasing demand for agricultural commodities, fast shrinkage in agricultural area, declining productivity of land, ever-increasing fragmentation, and the alarming stage of tenancy system. The climate change aggravates the problem further. In these circumstances, India is yet to make a steady growth in the per capita agricultural production (refer Table 16.3).

Table 16.3: Per Capita Agricultural Production Index (1991-2001=100)

Countries	*1994-96*	*1999-2001*	*2003*	*2004*	*2005*	*2006*	*2007*
India	96	100	99	96	99	103	108
China	83	100	106	113	116	119	120
Pakistan	97	100	98	103	105	105	107

Source: Food and Agricultural Statistical Report, 2009.

India is a front ranking country in agricultural production, particularly in perishable items but when it comes to the processing of such production it has recorded abysmally low performance. India produces 41 per cent of world's mango, 23 per cent of banana, 24 per cent of cashew nut, 10 per cent of onion, 30 per cent of cauliflower, and 36 per cent of green peas. It has 53 per cent of world's buffalo and 17 per cent of goat population. The food processing level in major countries are: USA—80 per cent, France—70 per cent, Malaysia—80 per cent, Thailand—30 per cent, India—1.3 per cent. India has on a macro level 15 agro-climatic zones and 128 micro agro-climatic zones. The government has set up 61 Agri-Export Zones. This approach in segmenting agricultural production and processing creates a conducive atmosphere in bringing total quality management in sustainable agricultural production, coupled with a proper capacity building in this vital sector. The recent data, both at micro and macro level shows gradual decline in the overall agricultural production. The average yield for most of the vital agricultural products is far below the world's highest. This indicates that a segmented approach in agricultural production is highly needed to make the agriculture more remunerative and a major contributor to GDP. Therefore, TQM approach in agricultural production is the need of the hour (*see* Table 16.4).

India possesses many credits in terms of various inputs for food production. India occupies first or second position in practically all items of mass consumption and in area of cultivation. It is pointed out that even though India has high stake in food processing inputs, its position is not showing any improvement. The widening gap needs to be urgently plugged to make the agricultural production more profitable (*see* Table 16.5).

Table 16.4: Average Yield and Yield Potential Unit: Tonnes per Hectare

Food Items	*India*	*World Highest (Country)*
Rice	3.03	9.71 (Egypt)
Wheat	2.69	8.89 (Namibia)
Total Cereals	2.39	10.41 (Ireland)
Pulses	0.60	5.14 (Barbados)
Edible Oil	0.25	4.29 (Peru)
Sugarcane	60.7	122.70 (Malaysia)

Source: Surabhi Mittal.

Table 16.5: Leading Exporters and Importers of Agro Products in 2006

Sl. No.	*Exporters*	*Billion Dollars*	*Importers*	*Billion Dollars*
1.	European Union	405.25	European Union	433.66
2.	Other European Countries	95.31	Other European Countries	123.72
3.	USA	92.66	USA	103.65
4.	Canada	44.23	Japan	65.62
5.	Brazil	39.53	China	51.65
6.	China	32.54	Canada	23.95
7.	Australia	22.18	Russian Federation	23.38
8.	Thailand	21.58	South Korea	18.58
9.	Argentina	21.33	Mexico	18.46
10.	Indonesia	18.32	Hong Kong China	18.46
11.	Russian Federation	17.06	Retained Imports	9.79
12.	Malaysia	15.57	Taipei, Chinese	9.67
13.	Mexico	14.69	Switzerland	8.85
14.	India	14.41	UAE	8.81
15.	New Zealand	13.24	Saudi Arabia	8.56
16.	Chile	11.49	Malaysia	8.50
			India	7.84

Source: WTO.

Food Market and ASEAN Partners

Major food producers in the world are concentrated in the Asian countries but the level of processing is high in European and American countries. Food shortage is rampant in African countries. This asymmetrical distribution of the world food market poses a vital problem in rising food prices and equitable reach of food products across the world. The level and intensity of waste in food items are higher (in certain cases in India it is 15 to 25 per cent of total production) in Asian countries due to the scant level of quality processing resulting in spiraling transaction cost and rise in the prices of food products. This results in the farmers getting less than average return. The overdependency upon agriculture by the population results in the delayed economic development or absence of sophistication in life and consequently poor quality of life index. This phenomenon can be addressed through proper institutional mechanism from the public, private and co-operative sector.

Why Sustainable Consumption

India poised to become a developed country expecting a GDP growth of 10 per cent annually. The rapid rise in consumption levels due to the increase in the disposable income aggravates food security of the nation. Given the negative growth of the agriculture and allied sector coupled with the increase in the projected growth of Indian food sector (9 %) creates a daunting task on policy-makers. "Long term food security demands that research in production technology of non-cereal food, through technology access to the poor small producers, should be promoted. Improvement in the quality of food items and reduction in transaction costs associated with their market access need to policy priorities."[3]

Table 16.6: Demand Supply Gap of Selected Food Items at Projected 9 per cent Growth of GDP

(Unit: Million Metric Tonnes)

Commodities	*2011*			*2021*			*2026*		
	D	*S*	*G*	*D*	*S*	*G*	*D*	*S*	*G*
Rice	94.4	95.7	**1.26**	96.8	105.8	**8.98**	102.1	111.2	**9.13**
Wheat	59.0	80.2	**21.21**	64.3	91.6	**27.33**	65.9	97.9	**32.04**
Total Cereals	188.5	209.7	**21.19**	245.1	242.2	**-2.94**	277.2	260.2	**-16.97**
Pulses	24.1	16.1	**-8.05**	42.5	17.6	**29.92**	57.7	18.4	**-39.31**
Edible Oil	16.8	10.1	**-6.66**	30.2	12.5	**17.68**	40.9	13.9	**-26.99**
Sugar	29.3	25.0	**-4.31**	65.7	26.0	**39.67**	100.7	26.6	**-74.13**

D = Demand, S = Supply, G = Gap

Source: Surabhi Mittal, Working paper, Demand Supply Trends and Projections of Food in India, March 2008, Indian Council for Research on International Economic Relations.

Issues to be Addressed

The success of agro processing and food processing in one country depends on vertical integration based on transaction cost analysis, performance of supply chain, buyer-seller partnering, relationship investment and channel partnering. Vertical integration is nothing but the forward and backward integration of connected activities to complete the value chain. This ensures the better price discovery, financial inclusion, social security, emergence of financial services etc. Food processing parks are major step in this direction but it is yet to be emerged in its true sense based on the transaction cost.

The food processing sector needs to be revamped from the low value addition to the high end consumption products. This requires an accurate planning and monitoring of the entire elements of the value chain. The selection of high yielding variety of seed, skilled workforce, cultivation based on economies of scale, application of judicious mix of organic and inorganic fertilizers, harvesting, grading, storage in cold chains, moving to the production, packaging, quality certification, selling and distribution. The biggest challenge is poor purchasing power of the rural population. Considering the increased incidence of wastage and higher storage costs a total revamping is needed in production, processing and consumption of agricultural commodities. "The argument that overall economic growth would drive income, and higher income will ensure food security has not worked so far, also because the higher food output has also come at a higher cost. Food security for the poor would have to make food available at an affordable price."[4]

Value Chain Analysis in Agro Processing and Food Processing

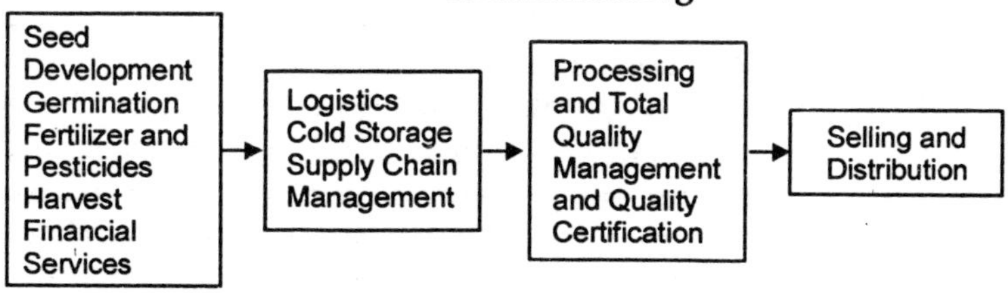

Transaction Costs in Food Processing Sector

India, even though the second largest agricultural producer in the world, there is an acute shortage of processors. The fast growing retail sector establishing the long term relationship with farmers will result in the faster growth of food processing sector. Now it is interesting to note that processing firms are non-existent or existing in a fragmented manner. Therefore, it is imperative to analyze the transaction cost in food processing industry. Lowering transaction cost is the basis of existence of any firm. The forward and backward integration as a strategy to bring in food processing industries in India will lead to cost leadership and competitive advantage to Indian companies in agro processing and food processing. This will have a cascading effect, eventually resulting in the economic well-being of agriculturists and farmers. 56.7 per cent of the total workforce of India is engaged in agriculture, 12.1 per cent in manufacturing and 31.2 per cent in service sector. This indicates the workforce sophistication should start from the agricultural sector so as to improve the purchasing power of the rural India. "As simple as cleaning, washing, sorting and packaging of fruits, vegetables and other farm products at the farm or village level can provide tremendous employment and also save huge physical losses of product that occur during storage, handling and transportation to the market."[5]

Institutional Mechanism—A Future Direction

The institutional mechanism at the government level should be strengthened. In the changed context, agriculture itself takes the form of industry. The issues, challenges, problems and prospects affecting the industry are more or less similar to agriculture. Therefore, all state governments must coordinate the activities under the Ministry of Agriculture and Food Processing. The local self-government must be encouraged to take up responsibility in agriculture production and processing. Venture capital on case by case basis should be provided. The co-operative sector mechanism should be effectively utilized. The major impediments to the growth of agricultural sector are the level of unorganized sector and the quality problem associated with it. This to a certain extent can be solved by forming co-operatives at Panchayat level and also by forming Self Help Groups. National and international quality assurance certification and good manufacturing practices can be taught to the processors and they should be insisted to abide by the same.

Total Quality Management for Sustainable Production and Consumption

The major impeding factor in the agricultural production and consumption is the quality issues associated with it. Strict adherence to quality is necessary to make the value addition process in agriculture. This moves at par with the value chain in agriculture and food processing. Therefore, farmers and consumers are linked and slowly the unorganized sector should be part of the quality food processing system. Total quality management in food processing is now confined to HACCP (Hazard Analysis and Critical Control Point) and ISO 22000. Even though India follows many quality standards like FPO, Agmark, etc., it is more focused on final products. But the quality process must start from the raw material selection till it finally reaches in the hands of the consumers. There is an urgent need for uniform quality code for all items of mass consumption including that which is produced through unorganized sector. HACCP certification is insisted in most of the exports. "Efforts are also needed to develop HACCP and other safety measures for each of the potential fresh/processed food items separately since single system are not universally acceptable."[6] The presence of private players in agricultural production and its final marketing creates quality related issue at the time of agricultural production till its final consumption. The quality takes a backseat when the country is not a major player in the international market, consequently the sector provides little scope for improvement. Therefore, TQM approach improves the overall quality, such as quality of capacity building, quality benchmarking in the final product, transaction cost reduction, value chain analysis, segmented approach like agro climatic zones, area coverage, technology use, crisis management, change in climatic conditions etc. This approach gives a cascading effect on the overall development of the country which gives fresh impetus to protection of farm land, responsible tourism, entrepreneurship development in agriculture and allied sector, environment protection.

The consumption pattern in India is quite different as compared to other countries in the world. The policy taken in this direction therefore should address the health and nutritional needs of the masses. This is quite impossible because of the poor purchasing power of the typical Indian consumer. Therefore, quality management should address price sensitivity of the consumer. The easiest way is to improve the abysmally low agriculture output per acre and also to reduce the higher incidence of wastage of agricultural produce. The storage cost of the agriculture should come down and bring technology to higher level of shelve value of the end product. This will bring about international competitiveness of India's agricultural products.

Conclusion

Total quality management in food processing industries is gaining momentum all over the world. Quality of the food for consumption depends on the level of sophistication in the agricultural sector in the country. Agro processing and food processing is highly interconnected. Total quality management in food processing sector cannot be possible without bringing total quality in agro processing. India has a supreme position in the world in agriculture, but the growth of the sector is extremely poor. The processing of food in India is still worse. Therefore, it is required to have an integrated view on agricultural sector based on value chain and transaction cost. This will bring in the required level of sophistication in this vital sector of the country so as to economically emancipate the major section of population of India. India being a signatory to the ASEAN agreement should regain its position in agricultural sector based on productivity and profitability by fully committing itself in total quality management. The present institutional mechanism needs to be strengthened to meet the challenges of bringing total quality management in agro processing and food processing. Therefore, it can be concluded that sustainable production and consumption in India requires an urgent attention of policy-makers because it is matter of bringing sophistication to the majority of Indians and the sustainable consumption needs to address the health and nutritional problems of the masses. TQM approach is expected to bring faster development in a time bound manner.

NOTES

1. M.S. Swaminathan, "Sustainable Food Security, *Yojana,*" October 2010, pp. 5-9.
2. NABARD Report, www.nabard.org/ruraleconomy/ecosenario.asp
3. Surabhi Mittal: Structural shift in demand for food: Projections for 2020. (Aug 2006), Indian Council for Research on International Economic Relations.
4. Shashaka Bhide, "Impediments to food security" *Business Line,* Wednesday June 16, 2010, Vol. 17, No. 166.
5. "Employment opportunities in farm and non-farm sectors through technological intervention with emphasis on primary value addition", *National Academy of Agricultural Sciences,* New Delhi, October 2006, p. 20.
6. "Sustainable agricultural export", *National Academy of Agricultural Sciences,* India, September 1999, p. 7.

REFERENCES

Boutlskid Michael A. (1998). *Food Supply Chain Management,* Blackwell Publishing.

India Food Report-2008.

Julie A. Caswell, Maury E. Bredahl and Neal H. Hooker (1998). *Review of Agricultural Economics,* Vol. 20, No. 2 (Autumn Winter), pp. 547-557.

Saul Klein, Gary L. Frazier and Victor J. Roth (1990). "A transaction cost analysis Model of Channel Integration in International Markets", *Journal of Marketing Research,* Vol. 27, No. 2 (May), pp. 196-208.

Yogendra Yadav, Sanjay Kumar (2006). "The food habits of a nation", *The Hindu,* Monday, August 14, p. 14.

17

A Study of Consumer's Attitude and Behaviour Towards a Sustainable Energy System in India

Suresh Malodia

Introduction

Today India is the second fastest growing economies in the world. To maintain this momentum, the biggest challenge is to cope up with the growing energy needs of the country. McKinsey has reported that, against 1.6 per cent growth rate of world energy requirements between 2006 to 2030, India's energy demand is expected to grow by 3.6 per cent per annum on average and will become four fold by 2020 and double by 2030. In order to cater to this demand India needs to increase the supply of energy by 3-4 times and generation capacity by approximately 6 times. The World is in a transition phase and energy is central to it. India has been responsible for almost 10% of the increase in global energy demand since 2000. Yet the increase in domestic energy production is far below than India's consumption needs. By 2040 more than 40% of primary energy supply will be imported, up from 32% in 2013.[1] Consequently, there is a large latent demand for energy services that needs to be fulfilled in order for people to have reasonable incomes and a decent quality of life.

Improving the energy efficiency meets the dual objectives of promoting sustainable development and of making the economy competitive.[2] Recognizing the formidable challenges of meeting the energy needs and providing adequate and varied energy of desired quality in a sustainable manner and at reasonable costs, improving efficiency have become important components of energy policy. In addition, the environmental and health burdens arising out of the use of hydrocarbons may also force mankind towards energy efficiency and clean energy systems.[3] Energy conservation has also assumed enhanced importance with a view to conserve depleting energy resources.

The challenge in creating a sustainable energy system for the future lies in two broad segments. The first is technology advancement and the second is conserving energy. In such a scenario, the cheapest source of power generation for India can be energy saving. Aristotle has rightly quoted, "Conservation is the first law of natural world". Wastage and inefficient use of energy will lead us to severe environmental consequences at regional as well as global level.

There is a huge scope of energy conservation in domestic sector. If we look at the sector-wise consumption pattern, the domestic sector accounts for 26 per cent of the total consumption and has a saving potential of 25-30 per cent. Approximately 20,000 MW (Megawatt—energy made available by means of energy conservation) energy can be made available through efficient use energy in power sector and 30-35 per cent savings are possible in oil and gas consumption. These savings will not only benefit the end-users in cost cutting but will also enhance our energy security.

Now the biggest challenge is how to motivate domestic consumers to conserve energy or say how we can alter the consumer behaviour towards energy conservation. The fact that energy consuming behaviour is largely based on our beliefs, values and attitude is well researched and proven over last 40 years. These behaviours are very complicated and thus difficult to change, however, these can be influenced by mass awareness, consumer education and guidance.

According to Martiskainen (2007), our energy consumption behaviours are based upon our daily routine and habits, viz., putting on lights every time even when not required, leaving the television, computers on standby mode are certain complicated behaviours which are difficult to change.

Conservation behaviours are significantly related to beliefs in material growth and comfort associated with it (Tashchian *et al.*, 1984) and therefore, promotional campaigns must be developed which takes into consideration specific behaviours associated to energy consumption. Incentives and disincentives policy for energy conservation should be framed after considering how people's behaviour are formed and influenced and how they could be influenced (Cunningham & Joseph, 1978).

A public opinion study was conducted by Future Foundation (2006) which revealed that, though people are aware about numerous benefits of energy conservation on environment as well as savings upon their own wallets but still there is a lack of knowledge about energy use and moreover, 40 per cent respondents said they have more important things to worry rather than analyzing and thinking how much energy they are using. Major challenges pertaining to energy conservation behaviour is how can we make use of energy and its impact more visible to the domestic consumers which can motivate them to conserve more for most of the time energy use in home is invisible (Abrahamse, *et al.*, 2005; Darby 2006).

Behaviour can, however, be influenced and in some cases we can also change it but in contrast to some other environmental issues behaviours *viz.* as green consumerism, waste reduction, the behaviour related to energy consumption needs more in-depth analysis due to multiplicity of behaviours involved (Hayes & Cone, 1977). Large number of internal and external factors *viz.* beliefs, values and attitude and group influence, culture, economic incentives etc. affect the usage of energy.

However, a lot of research has been conducted to assess public attitude towards sustainable energy consumption and a number of attitudinal/behavioural factors that influence the behaviour towards energy conservation have been identified. None of the studies provide clear evidence that what measures and which combination would be most effective in achieving sustainable energy consumption behaviour among end-users. Similarly, no study, so far has tried to assess the factors affecting energy consumption behaviour in terms of their ability to differentiate among the energy conservers against non-conservers.

Theoretical Framework

A number of studies have been conducted so far to assess public attitudes towards energy crisis and energy conservation. Several studies have used factor analysis to identify the underlying dimensions of attitudes towards energy conservation (Talarzyk & Omura, 1975; Barnaby & Reizenstein, 1976; Cunningham & Lopreato,

1977; Durand, 1979; Seligman *et al.*, 1979; Becker *et al.*, 1981 and Taschian *et al.*, 1984). A framework was presented by Milstein (1979) which was a synthesis of all past studies and he suggested three attitudinal factors i.e., cynicism, belief in material growth and belief in technology to be important in determining energy conservation. Tashchian *et al.* (1984) used confirmatory factor analysis to validate that these three scales actually represent three mutually exclusive attitudinal dimensions towards energy conservation.

Cynicism

Under this attitude, consumers do not consider energy conservation as a problem that concerns them, such consumers hold government to be responsible for supply of energy (Raaij & Verhallen, 1983). Cynics believe shortage of energy is not a real problem but a hype created by major oil companies and the government (Tashchian *et al.*, 1984). They consider energy crisis as a short-term condition due to political and economic manipulations for justifying the price increase. This type of a perception is a major barrier in developing favourable attitude towards energy conservation. Hence, the energy conservation campaigns carried out by government must address to this mindset make consumers perceive the energy problem as a real and serious issue. Underlying variables used to measure cynicism in various studies are blame and responsibility for energy crisis (Talarzyk & Omura, 1975); responsibility for the energy crisis, who has taken advantage of the energy problem, who has tried to solve the energy problem (Cunningham & Lopreato, 1977); perception of corporate advertising on energy issues (Durand 1979); energy crisis is a situation contrived by major oil companies to boost their profits, government serves interest of oil companies (Tashchian *et al.*, 1984).

Belief in Material Growth

With the growing per capita disposable income, the size of affluent society is growing at a very fast pace in India and place a high value on material comforts and conveniences. This has increased the demand for energy based living standards. According to Tashchian *et al.* (1984), the affluent segment of the society believes that material acquisitions distinguish the successful from the unsuccessful. Martiskainen (2007) in his report states that energy conservation demands for change in assets accumulation patterns and at this point majority of people choose comfort and convenience and consider energy conservation measures as working against their comfort and lifestyle. Comfort and Health explained the highest variance in a four factor solution which was 30 per cent in the study conducted by Seligman *et al.*, (1979). Attitude factors *viz.*, economic effort, comfort and health, effort and resultant savings, comfort and convenience indicate a general factor related to material growth (Talarzyk & Omura, 1975; Seligman *et al.*, 1979; Becker *et al.*, 1981).

Faith in Technology

People believe that science and technology in our country can take care of future energy needs. Nuclear energy is considered to be a very strong option by the people who have faith in technology and believe that technology would definitely be able to develop new sources of energy and provide more efficient ways to cater to the future energy needs. Therefore, those who believe in technology think energy conservation by individuals cannot help at all, new technology can only resolve the issue. The faith in technology attitude factor, optimism are related to faith in technology (Becker *et al.*, 1981; Durand 1979).

Objective

The objective of this chapter is to understand the behavioural dimensions of consumers towards energy conservation and assess the factors affecting attitude towards energy conservation using Discriminant analysis in order to identify the factors which differentiate between consumers who have a positive attitude towards energy conservation and those who have a negative attitude towards energy conservation. Study also intends to suggest a behavioural framework that can help in formulating positive attitude towards energy conservation among end-users. This will also help organization like Bureau of Energy Efficiency (BEE), Petroleum Conservation and Research Association (PCRA) and other central and state agencies in their Demand Side Management (DSM) programmes and achieve the goals of Energy Conservation Act, 2001 via more effective mass awareness programmes, consumer education and guidance programmes.

Methodology

The objective of the research work is to identify the difference between grey consumers (those who do not believe in energy conservation at all), grey-green consumers (those who moderately believe and practice energy conservation behaviour) and green consumers (those who have a highly favourable attitude towards energy conservation) on the basis of their attitude towards energy conservation. The three attitudinal factors were selected which Milstein (1979) has hypothesized and Tashchian *et al.* (1984) had validated using confirmatory factor analysis and by testing internal validity. These factors are cynicism, material growth, and faith in technology. Based on Tashchian's definition of these factors, a representative pool of 12 items was developed. Following items were included in the final scale.

Cynicism

- The energy crisis is a situation contrived by major oil companies to increase their profits.
- The consumer is helpless in attempting to reduce energy bills as the oil companies will continue to manipulate the supply and price of energy.
- Government serves the interest of private power corporations and does not look for real ways to solve our energy problem.
- Belief in Material Growth.
- Material growth makes for happier living.
- Growth in material consumption helps raise the level of civilization.
- I would rather be perfectly comfortable in my home (air conditioned—neither warm nor cold) than be slightly comfortable to conserve electricity.
- I have worked hard to get where I am and am entitled to the best things in life.
- People should heat and cool their homes to the most comfortable temperatures regardless of what the government says.

Faith in Technology

- Technological power does not possess the key to solve our energy problem.
- With every technological advance, the energy situation is getting more worse.
- Petroleum is the only source of energy, technology cannot find other significant energy sources.
- Energy conservation by individuals can help a lot, new technology alone cannot resolve the issue.

After this, to classify consumers between high and low conservation groups they were asked to rate their efforts in adopting each of the following conservation measures on three dimensions of energy conservation conserving the energy at home by reducing the wastage, conserving the energy by purchasing only energy efficient or say star rating appliances and replacing the inefficient and old appliance with more efficient appliances.

Sampling Design

The respondents in this study were selected by the use of systematic sampling procedure. The sample considered for this study included the residents of Dehradun, residing in western part i.e., Chakrata Road, GMS Road, Ballupur, Indranagar, Vasant Vihar, Rajendra Nagar. A team of 4 students from MBA Power Management 2009-2011 batch was selected for the interview and were given adequate training on survey methods. Each surveyor was assigned different areas and every fifth house was to be selected as a sample unit. Self-administered questionnaires were used to collect the data. The survey team visited the sample units personally and got the questionnaires filled in their presence. A total of 300 questionnaries were distributed i.e., 75 questionnaries per surveyor and 221 were found suitable for the analysis. In order to find sampling bias randomly 10 questionnaires were selected from the set of questionnaires of each surveyor, i.e., total 40 respondents were verified by calling at their numbers and talking to them. The profile of the respondents is shown in Table 17.1.

Table 17.1: Profile of Respondents

Male	101 (45.7%)
Female	120 (54.29%)
Age Group	
18-25 year	53 (23.98%)
25-35 years	62 (28.05%)
35-45 years	48 (21.72%)
45-55 years	42 (19%)
55 years or more	16 (7.23%)

The following are the results of estimating three groups Discriminant function. An examination of group (Table 17.2) indicates that belief in material growth appears to separate the groups more widely followed by cynicism and faith in technology. Both belief in material growth and cynicism have large standard deviations.

Table 17.2: Group Statistics

Energy Conserver			*Mean*	*Std. Deviation*	*Valid N (list-wise)*	
					Unweighted	*Weightec*
Dimensionon	Grey Consumers	Cynicism	12.36	4.805	76	76.000
		MatGrowth	24.29	4.357	76	76.000
		FaithTech	8.82	3.093	76	76.000
	Grey/Green Consumers	Cynicism	18.99	3.470	79	79.000
		MatGrowth	21.08	4.687	79	79.000
		FaithTech	10.25	3.152	79	79.000
	Green Consumers	Cynicism	15.58	4.110	66	66.000
		MatGrowth	15.45	3.767	66	66.000
		FaithTech	13.24	3.692	66	66.000
	Total	Cynicism	15.69	4.988	221	221.000
		MatGrowth	20.50	5.583	221	221.000
		FaithTech	10.65	3.748	221	221.000

The significance attached to univariate F ratios (Table 17.3) indicates that when the predictors are considered individually all three are significant in differentiating between the two groups.

Table 17.3: Tests of Equality of Group Means

	Wilks' Lambda	*F*	*df1*	*df2*	*Sig.*
Cynicism	.689	49.313	2	218	.000
MatGrwth	.592	75.119	2	218	.000
FaithTech	.770	32.615	2	218	.000

The pooled within group correlation matrix (Table 17.4) indicates very weak (negligible) correlations between variables indicating that multicollinearity will not be a concern.

Table 17.4: Pooled Within-Groups Matrices

		Cynicism	*MatGrwth*	*FaithTech*
Correlation	Cynicism	1.000	.080	.003
	MatGrowth	.080	1.000	.101
	FaithTech	.003	.101	1.000

Summary of Canonical Discriminant Functions

Because there are three groups, a maximum of two functions can be extracted. The eigen value (Table 17.5) associated with the first function is 1.225, and this function accounts for 77.3 per cent of the explained variance. Because the eigen value is large, the first function is likely to be more superior. The second function has a eigen value of .360 and accounts for only 22.7 per cent of explained variance.

Table 17.5: Eigen values

Function	*Eigen value*	*Per cent of Variance*	*Cumulative per cent*	*Canonical Correlation*
1	1.225[a]	77.3	77.3	.742
2	.360[a]	22.7	100.0	.515

The value of wilks' λ (Table 17.6) is 0.330. This transforms to a chi-square of 240.316 with 6 degree of freedom which is significant beyond 0.05 levels. Thus, the two functions together significantly discriminate between the three groups.

The standardized canonical discriminant function coefficients (Table 17.7) indicate a large negative coefficient for Belief in material growth for function 1 whereas function 2 has relatively

Table 17.6: Wilks' Lambda (λ)

Test of Function(s)		*Wilks' Lambda*	*Chi-square*	*df*	*Sig.*
Dimension	1 through 2	.330	240.316	6	.000
	2	.735	66.739	2	.000

Table 17.7: Standardized Canonical Discriminant Function Co-efficients

	Function	
	1	2
Cynicism	.395	.922
MatGrowth	–.821	.245
FaithTech	.560	–.261

larger coefficients for cynicism and faith in technology. A similar conclusion is reached by an examination of the structure matrix (Table 17.8).

Pooled within-groups correlations between discriminating variables and standardized canonical discriminant functions Variables ordered by absolute size of correlation within function. (*see* Table 17.9).

Table 17.8: Structure Matrix

	Function	
MatGrowth	–.733*	.293
FaithTech	.478*	–.234
Cynicism	.330	.941*

Table 17.9: Functions at Group Centroids

Energy Conserver	*Function*	
	1	2
Grey Consumers	-1.349	-.378
Grey/Green Consumers	.137	.796
Green Consumers	1.389	-.517

Note: Unstandardized canonical discriminant functions evaluated at group means.

Classification Statistics

The classification results based on the analysis sample indicate that (64 + 55 + 51)/221 i.e. 76.92 per cent cases are correctly classified. Leave one out cross-validation correctly classifies only (63 + 54 + 51)/221 = 76.02 per cent of the cases (*see* Table 17.10).

Table 17.10: Classification Results[b, c]

		Energy Conserver	*Predicted Group Membership*			
			Grey Consumers	Grey/Green Consumers	Green Consumers	Total
Original	Count	Grey Consumers	64	9	3	76
		Grey/Green Consumers	13	55	11	79
		Green Consumers	4	11	51	66
	Per cent	Grey Consumers	84.2	11.8	3.9	100.0
		Grey/Green Consumers	16.5	69.6	13.9	100.0
		Green Consumers	6.1	16.7	77.3	100.0
Cross-validated[a]	Count	Grey Consumers	63	10	3	76
		Grey/Green Consumers	13	54	12	79
		Green Consumers	4	11	51	66
	Per cent	Grey Consumers	82.9	13.2	3.9	100.0
		Grey/Green Consumers	16.5	68.4	15.2	100.0
		Green Consumers	6.1	16.7	77.3	100.0

Findings and Suggestions

The study examined how respondents who show high interest in energy conservation differ in their attitude towards energy conservation from those who show low interest. Group results suggests that the three groups are different primarily in terms of Belief in Material Growth, then Faith in Technology is the second most important factor, whereas, cynicism is not related to conservation behaviour. The analysis shows that people who are highly cynical towards energy crisis need not necessarily have a negative attitude towards energy conservation. Those who believe in material growth were found to be reluctant towards conserving energy. Similarly, those who have faith in technology are likely to give little less emphasis on energy conservation.

Knowing what factors influence energy consuming behaviours, the question that remains unanswered is what promotional methods and measures will help in changing the energy consuming behaviour favourably. Mc Makin *et al.* (2002) pointed out three conditions under which people are likely to adopt energy conserving habits:

- People will conserve energy if they see self benefit,
- Energy use and savings must be visible objectively, and
- Personalized information is helpful if presented in a clear way.

Tashchian *et al.* (1984) states that promotional campaigns intended to change the attitude will be discounted by public and hence, does not result in desired conservation behaviour. Small amount of fear can have a positive effect in encouraging behaviour change. However, too much fear can have the opposite effect, causing no behaviour change and undue worry and stress for people receiving such information (Uzzell *et al.*, 2006, p. 25). Hence, promotion will be most effective when it is aimed at enhancing the attitude rather than trying to change it. Those who believe in material growth may hence be motivated to replace their energy appliances with modern appliances which are more eco-friendly, comfortable and will fulfil their status needs. Similarly, those who do not believe in material growth and fall in the third category of green consumers may be motivated to adopt behaviour which will conserve energy in their daily use.

Demarketing and disincentives for consumption rather than marketing for conservation or say incentives for conservation is also found helpful in reducing the usage of energy resources, specially in petroleum related products. Increased cost of road tax, parking will encourage use of public transport whereas at the same time incentive scheme for car pooling and use of public transportation does not work that much effectively.

To sum up, this investigation finds that there is a significant difference between those who show interest in energy conservation and those who just do not consider energy conservation important in terms of belief in material growth and faith in technology. The major limitation of this study is that it is based on a very small sample from a single city and hence the results cannot be generalized. However, the results are consistent with the previous studies conducted.

NOTES

1. UNDP India, Eight Goals for 2015, website: undp in.org/content/india/home/post-2015/mdgoview.html.
2. National Electricity, The Gazette of India, Ministry of Power, Government of India. Website: powermin.nic in en/content national-electrcity-policy.
3. Hydro Power Policy 2008. Ministry of Power, Government of India, pp. 1-80.

REFERENCES

Abrahamse W., Steg L., Vlek C. and Rothengatter T. (2005). "A Review of Intervention Studies Aimed at Household Energy Conservation." *Journal of Environmental Pyschology,* Vol. 25, pp. 273-291.

Barnaby, David J. and Reizenstein C. Richard (1976). "Attitude Towards Energy Consumption: Segmenting the Gasoline Market." *Advances in Consumer Research,* Vol. 3, pp. 246-251.

Becker B.W., Brown Daniel and Scharif B. Phillip (1976). "Behaviour of Car Owners During Gasoline Shortage." *Traffic Quarterly,* Vol. 3, pp. 469-483.

Cunningham William H. and Lopreato Cook Sally (1977). *Energy Use and Conservation Incentives: A Study of the South-western States.* New York: Praeger Publishers.

Cunningham, W.H. and B. Joseph (1978). "Energy Conservation, Price Increases and Payback Periods." *Advances in Consumer Research,* Vol. 5 Association for Consumer Research, pp. 201-205.

Darby S. (2006). *The Effectiveness of Feedback on Energy Consumption—A Review of Defra of the Literature on Metering, Billing and Direct Displays.* Envirnomental Change Institute, Oxford University.

Durand Richard M. (1979). *A Study of Alabama Consumer Attitudes Towards Energy Crisis.* School of Mines and Energy Department: The University of Alabama (August).

Future Foundation (2006). *Energy Efficiency—Public Attitude, Private Action,* Logica CMG.

Hayes S.C. and Cone J.D. (1977). "Reducing Residential Electrical Energy Use: Payment, Information and Feedback." *Journal of Applied Behaviour Analysis,* Vol. 10: 425-435.

Martiskainen Mari (2007). *Affecting Consumer Behaviour on Energy Demand;* Final report to EdF Energy, Sussex Energy Group, University of Sussex.

McMakin A.H., Malone E.L. and Lundgren R.E. (2002)." Motivating Residents to Conserve Energy Without Financial Incentives. *Environment and Behaviour,"* Vol. 34: 848-863.

Milstein Jeffrey S. (1979). "The Conserver Society? Consumers Attitude and Behaviour Regarding Energy Conservation." *The Conserver Society,* pp. 43-50.

Raaij Van W.F. and Verhallen T.M.M. (1983). "Patterns of Residential Energy Behaviour." *Journal of Economic Psychology,* Vol. 4.

Seligman Clive, Kriss M., Darley M. John, Faxio H. Russell, Becker J. Lawrence and Pryor J.B. (1979). "Predicting Summer Energy Consumption from Homeowners Attitudes." *Journal of Applied Social Psychology,* Vol. 9: 70-90.

Talarzyk W. Wayne and Omura S. Glenn (1975). *Consumer Attitude Toward and Perceptions of the Energy Crisis;* American Marketing Association Combined Proceedings, pp. 316-322.

Tashchian Armen, Slama E. Mark, and Tashchian O. Roobian (1984). "Measuring Attitudes Towards Energy Conservation: Cynicism, Belief in Material Growth, and Faith in Technology." *Journal of Public Policy and Marketing,* Vol. 3: 134-148.

Uzzell D., Muckle R., Jackson T., Ogden J., Barnett J., Gatersleben B., Hegarty P. and Papathanasopoulou E. (2006). *Choice Matters, Alternative Approaches to Encourage Sustainable Consumption and Production.* Department of the Environment Food and Rural Affairs.

18

The Clean Development Mechanism (CDM) and Carbon Trading

Anirudh Prasad Singh and *Jyoti Singh*

Introduction

The Carbon trading is one of the fastest growing financial markets in the world. It is the most visible result of early regulatory efforts to mitigate climate change, and grew out of the Kyoto Protocol, which was adopted in 1997. The Clean Development Mechanism (CDM) allows emission-reduction projects in developing countries to earn Certified Emission Reduction (CER) credits, each equivalent to one tonne of CO_2. These CERs can be traded and sold, and used by industrialized countries to meet a part of their emission reduction targets under the Kyoto Protocol. The mechanism stimulates sustainable development and emission reductions, while giving industrialized countries some flexibility in how they meet their emission reduction limitation targets. The CDM is the main source of income for the United Nations Convention Framework on Climate Change (UNFCCC), which was established to finance adaptation projects and programmes in developing countries Parties who are part of the Kyoto Protocol.

The central feature of the Kyoto Protocol[2] was its requirement that countries limit or reduce their greenhouse gas emissions. By setting such targets, emission reductions took on economic value. To help countries meet their emission targets, and to encourage the private sector and developing countries to contribute to emission reduction efforts. Negotiation of the Protocol included three market-based mechanisms (i) emissions trading, (ii) the clean development mechanism (CDM)[3] and joint implementation. At the 18th Conference of the Parties[4] (COP 18), hold in Doha, Qatar in 2012, delegates agreed to extend the Kyoto Protocol until 2020. The new treaty, planned for implementation in 2020, would fully replace the Kyoto Protocol. But with the adoption of the Paris Agreement, in 2015,[5] the instrumental mechanism had been changed because it was a separate instrument under the UNFCCC rather than an amendment of the Kyoto Protocol. Joint Implementation Supervisory Committee (JISC)[6] took place its forty-second meeting, for the second time in a virtual environment, building on its experience in last year, 2019. Pending the Doha Amendment[7] into force, work pertaining to Joint Implementation was minimal. India offers a large potential for CDM because of its inherent dependence on fossil fuels for development. So countries with relatively low abatement and transaction costs like India are a major attraction for CDM projects.[8]

The market is emerging strongly despite various global factors, according to the World Bank. Regulation that caps greenhouse gas emissions has spawned an emerging carbon trade that was valued at US $64 billion (€47 billion) in 2007. For the third consecutive year, China was the world leader in CDM supply with a 73 per cent market share in terms of 2007 transacted volume. Brazil and India, at 6 per cent market share each, transacted the highest volumes after China. India is the fourth largest emitter of greenhouse gases in the world in absolute terms. But its per capita emission of 1.2 tons per person, per year is much lower than the West's figure of 20 tons, or than the global average of 8 tons. If India has to realize its ambitions of economic growth and take large sections of its population out of the low income trap, it must grow. That means greenhouse gas emission reductions will continue to be one of the country's greatest challenges.

Climate change and global warming is happening all around us. Soon it will start to shape every aspect of our lives in ways that we might not comprehend: from how we travel and what products we buy, to where we live and how we work. Hence, there is a concern that the use of non-renewable fuels and other human activities are increasing greenhouse gases in the atmosphere, contributing to global warming. To avoid this a new currency is emerging in world markets. Unlike the dollars, Euros and Yen that trade for tangible goods and human services, this new money exchanges for pollution—particularly emissions of carbon dioxide, which are caused by burning fossil fuels, are the leading cause of global climate change. Carbon credits, as they are called, are poised to transform the world energy system and thus the world economy.

Carbon credits as currency allow companies and individuals to compensate for their carbon emissions by either reducing carbon-dioxide output directly or offsetting their own output. Carbon Credits originate from the United Nation's Clean Development Mechanism (under the Kyoto Protocol) that allows a finite number of carbon credits to be bought and sold. It is vitally important for people to limit their personal impact on the environment and buy carbon credits to offset what they cannot reduce. The primary goal is to reduce emission of green house gases. By permitting allowances to be bought and sold, an operator can seek out the most cost-effective way of reducing its emissions, either by investing in cleaner machinery and practices or by purchasing emissions from another operator who already has excess capacity. Carbon trading under the Clean Development Mechanism (CDM) is big business on the open market. Projects such as developing renewable energy, improving polluting industries, and planting carbon sinks are being funded by carbon credits. Companies and individuals are becoming carbon neutral by reducing their emissions and offsetting. Many types of activities can generate carbon offsets. Renewable energy such as the wind farms, installations of solar panels, small hydro turbines, geothermal energy, and biomass energy can all create carbon offsets by displacing fossil fuels.

Developed countries have to spend nearly \$300-500 for every ton reduction in CO_2, against \$10-\$25 by developing countries. India's Green House Gas emission is below the target and so, it is entitled to sell surplus credits to developed countries. India is considered to claim about 31 per cent of the total world carbon trade, which can give \$25 billion by 2010. This is what makes trading in carbon credits such a great business opportunity.

India has emerged as the dark horse in this race as more than 200 Indian entities have applied for registering their CDM Project for availing carbon credits. Indian companies can have higher incomes from carbon credits than their core business. The carbon credit market \$25 billion last year and is growing at tremendous space, and there is a demand to reduce 1 billion ton of carbon emissions in the world, so that threats like global warming could be dealt with.

Wal-Mart, Dell, GE are all going GREEN and reducing Green House Gas and also purchasing carbon credits. These companies are bolstering their brand name and consumer confidence in their products.

Global Warming

Rapid industrialization and development as well as the non-sustainable consumption of resources and belching of pollutants in atmosphere, exceeded the absorption/purification capacity of the natural resources which resulted in global warming. Scientists have found the Earth's temperature increased by more than 0.8 degrees centigrade in the last 100 years. This rise is going to increase between 1.4 to 5.8 degrees centigrade by 2100 because of the human impact on the climate, unless some drastic change is made to check this global warming.

Initiative was taken by UN and its agencies to curb greenhouse gas emissions to check global warming. Supported by the first scientific assessment of the intergovernmental panel on climate change (IPCC, 1990), the UNFCCC set the objectives of the convention to stabilize greenhouse gas concentrations in the atmosphere to a safer level. It finally resulted in the Kyoto Protocol to the convention on climate change.

The Kyoto Protocol requires the developed countries to bring down their emissions by changing their lifestyle and developing new technologies to reduce their emissions. It requires the developed countries to reduce their Green House Gas emissions by at least 5.2 per cent of the 1990 emission levels during the period 2008-2012. The developing countries were exempted from these reductions so that their development was not checked.

At some places, the control of Green House Gas is very costly as compared to other favourable conditions. While the marginal abatement cost at the EU level is 54.3/t CO_2, Germany achieves its target with a cost of 13.5/t CO_2, at the corresponding margin; the corresponding figure for the Netherlands is 150.7/t CO_2.

Hence, three market-based mechanisms have been incorporated. These are:

1. Joint implementation among developed countries,
2. Clean development mechanism between a developed country and a developing country, and
3. Emission trading among developed countries.

Types of Emission Trading Programmes

Basically there are three types of emission trading programmes: 'cap and trade', 'baseline and credit', and 'offset'.

In a 'cap and trade' programme, the regulator establishes an overall limit on emission, the 'emissions cap', the total amount of a pollutant that the participants in the programme are allowed to emit in a given period (e.g. emission of a number of tonnes of the pollutant per year).

Baseline and credit or offset programmes typically define a baseline that varies with the output, and issue credits at the end of the period if a participant's actual emissions are below its baseline.

The participants in a baseline and credit (or 'averaging') programme have to 'earn ' credits before they can begin trading.

Offset programmes are used to compensate for (i.e. offset) the additional emissions from a new source or expansion of an existing one. Under such schemes, those responsible for the new or expanding source, purchase credits equal to emission reductions achieved by the existing sources.

Allocation of Allowances

Allocation of emission trading allowances is a foremost issue. Each industry needs to be allotted certain emission allowances which they can consume and if they increase their emission from allotted emissions, they can be penalized or they have to purchase emission allowances from industries consuming fewer amount of allowances.

Mainly three methods of allotment of allowances are there, but they can be adopted either in isolation or in combination. These methods are: (i) Auction: This alternative would involve the Member State auctioning the allowances initially. (ii) Grandfather: Under this alternative, allocations would be provided to sources based upon historical information period. (iii) Update: This alternative involves allocating to sources based upon information that is updated over time.

Monitoring, Reporting or Verification

Monitoring mechanism is one of the most critical components of the emission trading scheme, including the reporting system and verification of the reports and emission by regulating authorities. Generally, two types of monitoring, reporting and verification are required (a) at the emission source level, and (b) at member state level.

For monitoring and measurement of Green House Gases at

source, standardized methods need to be developed. The practice being observed is the calculation of emissions based on fuel and raw materials being used and the processing being done. Direct monitoring is a more reliable method. Since it is costly and methods may vary, it is not much in use. For a uniform measurement all over, it is important that the monitoring system specifies one approach or the other, because the two approaches can lead to different numbers, which would provide emission sources with an opportunity to game the system in search of the best result. For a downstream system, it is desirable that all covered emission sources present their calculation of emissions, including the fuel combusted as well as emission factors.

Carbon Transactions

Carbon transaction may be defined as the purchase contract whereby one party pays another party in exchange for a given quantity of Green House Gas (GHG) emission 'credits' that the buyer can use to meet its objectives *vis-à-vis* climate mitigation. Payments are made in various forms, such as cash, equity, debt or in kind contributions, i.e. providing technologies to abate GHG emissions. Analysis, however, is restricted to transactions where carbon credits are effectively measured and transferred.

There are two main categories of carbon transactions:

(i) Project-based transactions, and
(ii) Allowance-based transactions.

The Kyoto Protocol defines that developed countries can reduce GHG emissions by contributing to emission reducing projects in developing countries like India.

Important greenhouse gases are: carbon dioxide (CO_2), methane (CH_4), nitrous oxide (N_2O), hydrofluorocarbons (HFC), perfluorocarbons (PFC), and sulfur hexafluoride (SF_6). Water vapour is also an important greenhouse gas, but since humans do not generally have a direct affect on water vapour concentration in the atmosphere, it is not included in this chapter. Because each greenhouse gas traps different amounts of heat and stays in the atmosphere for different lengths of time, studies use measures of global warming potential (GWP) to compare between gases. Carbon dioxide is used as the benchmark, so all other gases are measured in carbon dioxide equivalence (CO_2e).

Table 18.1: The Global Warming Potential of Six Major Greenhouse Gases (This Measure Takes into Account the Heat Trapping Abilities and the Time the Gas Stays in the Atmosphere (IPCC 2001a, 2001b))

Gas	*Global Warming Potential*	*Atmospheric Life (years)*
CO_2	1	5 to 200
CH_4	21	12
N_2O	310	114
HFC	140 to 11,700	1.4 to 260
PFC	6,500 to 9,200	10,000 to 50,000+
SF_6	23,900	3200

CDM (Clean Development Mechanism)

Developed countries can implement projects that reduce emissions or remove carbon from the atmosphere in other developing countries in lieu of CERs (Certified Emission Reductions). These CERs can be used to meet the emission targets. The protocol stresses that such projects are to assist the developing countries host parties in achieving sustainable development. Further, the protocol refrains developed countries from using CERs generated out of nuclear facilities to meet the targets.

Emissions Trading

The protocol provides that developed countries can acquire units

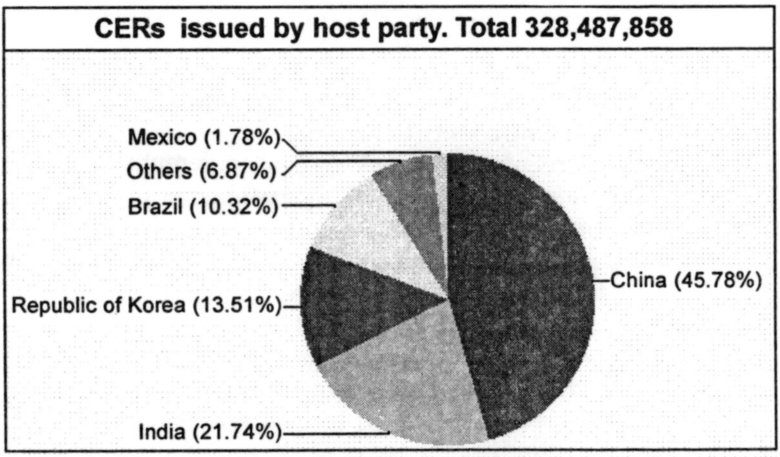

Source: http://cdm.unfccc.int(c)10.09.2009 14:55

from other developing parties and use them towards meeting their emissions targets, but must be prepared to transfer the units when they do not require them for compliance. This enables developed countries to make use of low cost opportunities to reduce emissions. Only some developed countries with emission limitation and reduction commitments specified in the protocol can engage in such an activity.

The latest analysis by New Carbon Finance shows that in spite of the global recession, the volume of trading in the global carbon market in the first quarter of 2009 grew by 37 per cent compared to the fourth quarter of 2008.

India CDM Methodology

In India, the Designated National Authority (DNA) is hosted by the Ministry of Environment and Forests (MoEF). In addition to the DNA, India has many state-level nodal agencies promoting and facilitating CDM-projects in their area. These organizations can be of assistance in setting up contacts with public organizations to arrange CDM projects, or in approaching a larger number of small-scale possible project proponents. They can act also as bundling agencies, i.e. combine a number of small-scale CDM projects and handle financial management for bundled projects.

Indian Scenario—Favouring Points

a. India—high potential of carbon credits
b. India can capture 10 per cent of Global CDM market
c. Annual revenue estimated range from US$10 million to 330 million
d. Wide spectrum of projects with different sizes
e. Vast technical human resource
f. Strong industrial base
g. Dynamic, transparent and speedy processing by Indian DNA (NCDMA) for host country approval
h. MoU Signed between MoP and GTZ (Oct. 2006)—Indo German Energy Programme (IGEP)
    - Baseline CO_2 Emissions from Power Sector already in place—first CDM country

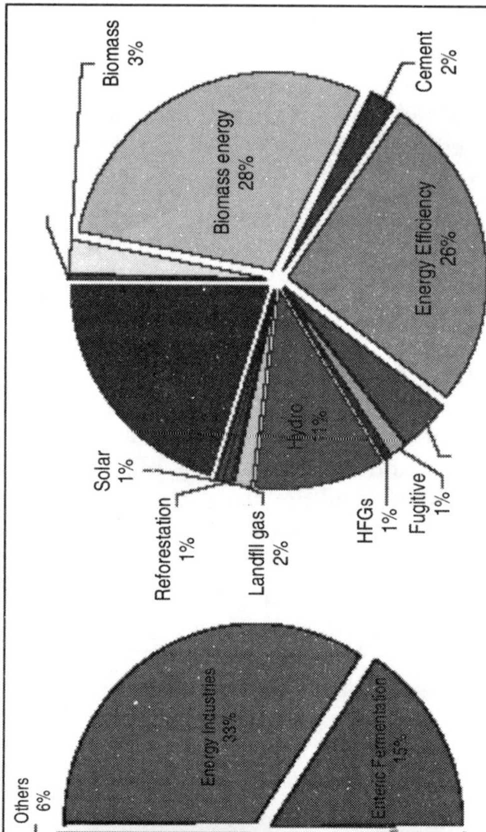

Emissions Break-up Sectorwise: India
Source: India's National Communication to UNFCCC, MoEF - 2004

"UNEP Risoe, India CDM/JI Pipeline Analysis and Database, June 11th 2008"

- Improvement in EE
- CDM in Power Sector.

CDM Potential for India

Sector-wise break-up: Investment done in host country approved project as on 2nd July, 2007. So far, India concentrated mainly on renewable energy (biomass, wind power, etc.)/waste heat recovery projects which generate much less Certified Emission Reduction (CER) as compared to HFC23 projects.

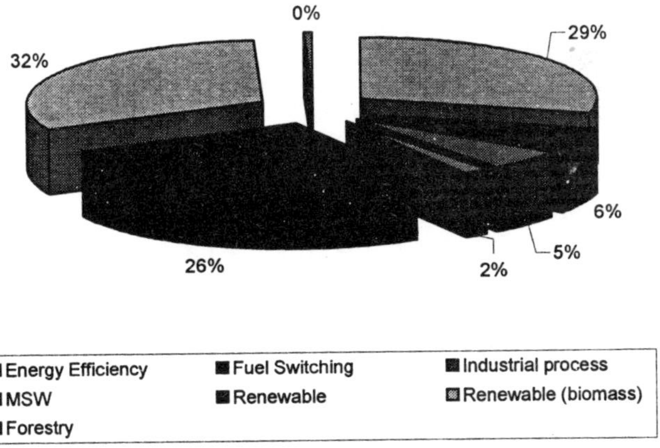

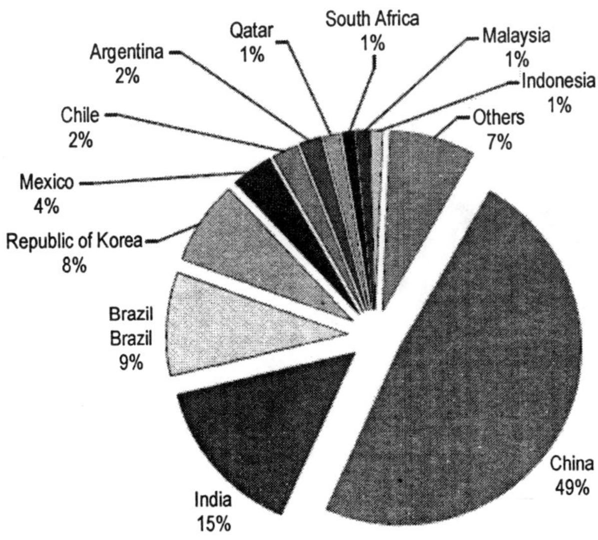

Policies and Way Ahead

Greenhouse gas abatement policy design is exceedingly difficult because GHG emissions result from nearly all modern human activities. It involves every sector of the economy as well as habits and choices of individuals. Economics is more than just a study of business; it is the science which studies human behaviour as a relationship between aspirations and the scarce means to reach those goals. Individuals make decisions everyday that influence the amount of greenhouse gases that enter the atmosphere. If a stable climate is one objective among the many to which society aspires, then economics is a tool, well-suited to understand how those decisions are made and how efficient and effective outcomes can be reached.

Indian Forum

India is a Party to the United Nations Framework Convention on Climate Change (UNFCCC) and the objective of the Convention is to achieve stabilization of greenhouse gas concentrations in the atmosphere at a level that would prevent dangerous anthropogenic interference with the climate system.

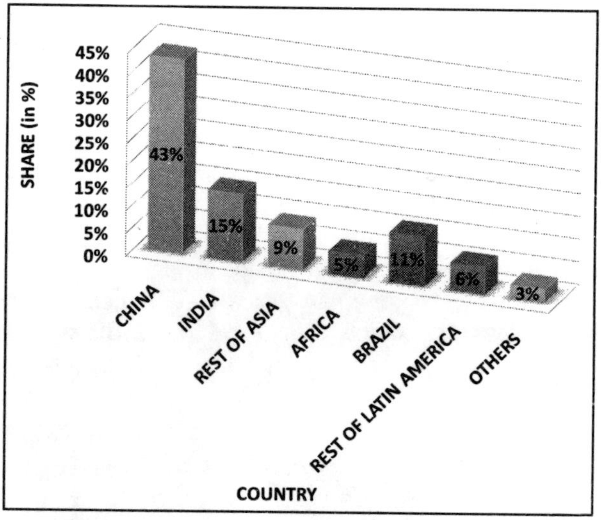

Fig. A: Expected Average Annual CERs from Registered Projects by Host Party.

Source: www.ficci.com/press/287/FICCI_CDM_Delegation_to_UK.doc

FIG. A—Inference: "Out of the total projects registered at UNFCCC, only China is ahead of India in generating CERs and got the highest share of 43 per cent followed by India (15%; 21,239,372 CERs)."

Out of the total 633 projects registered at the UNFCCC, 220 projects (34.76%) are from India that are expected to generate an annual volume of 21,239,372 CERs.

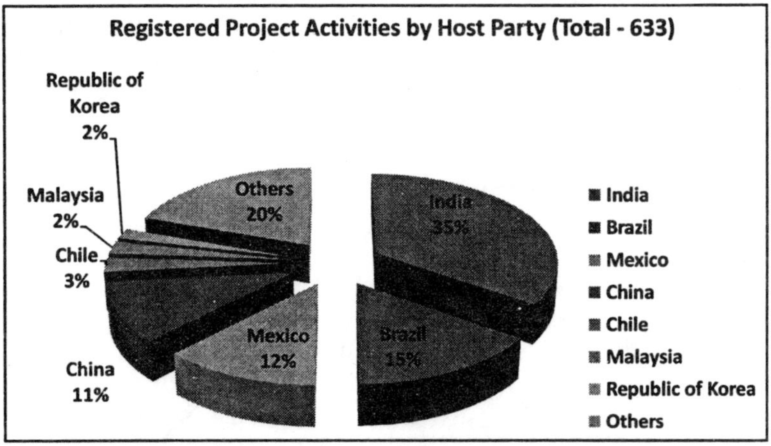

Fig. B: Registered Project Activities by Host Party.

Source: www.ficci.com/press/287/FICCI_CDM_Delegation_to_UK.doc

Fig. B: Inference: "India tops the chart in registering its projects at the global level. Out of 633 projects, India's share include 221 projects followed by Brazil (94 projects) which is far behind India. The closest rival, i.e., China has got only 70 projects registered."

Fig. A and Fig. B: Inference: "From the above figures, we can say that China is able to generate more of CERs, i.e., 43 per cent of the total with just 70 projects registered and India with highest registered projects numbering more than 220 is able to generate only 15 per cent of the total CERs, i.e., 22 million as against of 58 million of China. If we analyse it, we can say that India has registered thri:e the projects as of China but receiving less than one-third of CERs as compared to China."

Financing Sources

The financing patterns of CDM projects in India vary widely.

Central and state government financing institutions give financing support to CDM projects. So far there have not been CDM projects totally financed by Indian banks, but they are starting to show an interest. The State Bank of India, ICICI, IDBI and Yes Bank are to name a few. Banks do not act as outright CER buyers but have begun to offer specific financing for CDM projects. In return for a certain share of CERs, they offer lower interest rates on the project financing. Future developments might include loan schemes including carbon (future CER) collaterals. Financial institutions such as IREDA, IDFC and PFC/REC have been active in renewable energy financing and might become increasingly active in CDM project financing.

CDM Financing Instruments

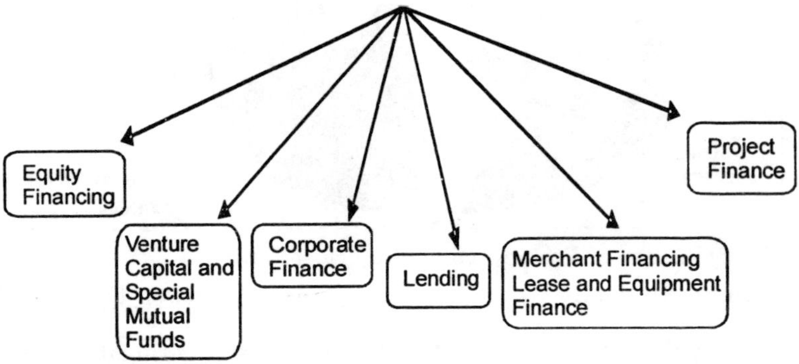

Fig. C: CDM Financing Instruments.

Conclusion

The above discussion clearly brings out that in recent years, significant global governmental involvement in the stabilization of the concentration of atmospheric green home gases has facilitated tremendous growth in the market for carbon emissions trading. India offers a large potential for CDM because of its inherent dependence on fossil fuels for development. So countries with relatively low abatement and transaction costs like India are a major attraction for CDM projects.

NOTES

1. Website: cdm.unfccc.int.
2. Website: unfcc.int./kyoto-protocol.
3. Website: cdm.unfcc.int.
4. Conference of the parties serving as the meeting of the parties to the Kyoto Protocol (CMP), Website: unfccc.int/process/bodies/ supreme bodies/conference of the parties serving as the meeting of the parties to the Kyoto Protocol.
5. Paris Agreement; United Nations climate change, Website: unfcc.int/process and meetings/the Paris agreement/what is the paris agreement.
6. UNFCCC joint implementation.
7. The Doha Agreement, Website unfccc.int/process/the Kyoto protocol; Doha Agreement 8 December, 2012.

REFERENCES

Clean Development Project Opportunities in India, TERI New Delhi, January 2001.

http://carbonfinance.org Router.cfm?Page=FAQ&ItemID=24677#2

http://cdm.unfccc.int/

http://cdmindia.nic.in/cdm_india.htm

http://economictimes.indiatimes.com/articleshow/1212812.

http://economictimes.indiatimes.com/articleshow/1937685.cms

http://unfccc.int/

http://www.cdmwatch.org

http://www.cliffordchance.com

http://www.ficci-cdm.biz/potential.asp

http://www.icai.org/*resource*_file/102641496-1500.pdf

http://www.indiainfoline.com

http://www.scribd.com/doc/8746421/Carbon-Credits-Implications-in-India-as-Financing-Instrument-Kalyan-Teja

http://www.srf-limited.com/

http://www.wikipedia.org/wiki/clean_development_mechanism

Kyoto Protocol to the United Nations Framework Convention on Climate Change, United Nations, 1998.

The Kyoto Protocol, http://unfccc.int/kyoto_protocol/items/2830.php, www.ieta.org

19

Groundwater Resource Depletion and Need for Development of Sustainable Communities

K.J. Anandha Kumar and *R.K. Mall*

Introduction

Water is the most wonderful gift of nature not only in making our planet beautiful, but also a dire necessity for the survival of the mankind. Water is key to sustainable development, crucial to its social, economic and environmental dimensions. Water plays a vital role in relation to human health, livelihood, economic growth helping to combat poverty as well as sustenance of ecosystems.

Groundwater has played a vital role in fulfilling the demands of domestic, agricultural and industrial sectors. Indiscriminate development and unscientific management of groundwater resource has led to problems of groundwater depletion and pollution.

To develop India's agricultural base and meet the growing water requirement of the domestic and industrial sector, the use of groundwater was initiated in 1950s as it was available at the place of requirement and cheaper. Coupled with this initiation, subsidy on the power for irrigation use has promoted the unplanned extensive use of groundwater. This has ultimately resulted in long term damage in some places where groundwater levels have gone down. The natural phenomenon like recurrent drought-flood-cyclone syndrome, tidal intrusions in the vicinity of the sea have restricted the availability of surface water round the year, putting stress on the groundwater resources. This has also led to salinity problems in the coastal areas. The extensive use of surface water available at the subsidized rate in the irrigation command on the other hand have also deteriorated the soil and has created waterlogging in the command areas, especially in the upper reaches.

The experience in the past three decades has shown that there is need to involve the community in management of groundwater resources, for its sustainable development. The experience of Central Ground Water Authority constituted under the Environment (Protection) Act, 1986, in 1997, to regulate indiscriminate use of groundwater, and the experience of other countries where the groundwater regulations are already in place show that it is essential to involve the communities in management of groundwater.

Water Demand

The total water use assessed in 2007 is 634 billion Cubic metre (BCM). The projected demand of water for 2011 at that time was 813 BCM and demand projected for 2025 and 2050 was 1093 BCM and 1447 BCM respectively (Planning Commission, 2007). The average annual rainfall of the country is 1170 mm, including 4000 BCM of snowfall. Out of this rainfall only 1869 BCM appears as average annual potential flow in rivers. However, the utilizable water is estimated to be only 1123 BCM, out of which 690 BCM is from surface water and 433 BCM is from groundwater. Thus, the demand will outstrip availability in another 35 to 40 years if appropriate measures are not taken which will include involvement of communities who will be the best managers of groundwater, the invisible resource. The total groundwater draft as per the last estimation (2004) is 230.6 BCM (Chatterjee Rana, 2009). It is estimated that there are 19 million wells in the country, out of which 16 million are in use and are drawing about 231 BCM, 213 BCM for irrigation and 18 BCM for domestic and industrial purpose. Further, it is estimated that by the year 2025, the demand for domestic and industrial uses is projected to rise to 29 BCM from the level of 18 BCM in 2004 (Planning Commission, 2007).

Groundwater is a replenishable and considered to be a finite resource. Rainfall is the principle source of its recharge, though in some areas canal seepage and return flow from irrigation also contribute significantly to the groundwater recharge. Groundwater resources comprises of two parts namely dynamic, in the zone of water table fluctuation and static resources, blow this zone, which usually remains perennially saturated. As per the National Water Policy, 2002, the dynamic groundwater resource is essentially the exploitable quantity of groundwater, which is recharged annually, and is also termed as replenishable groundwater resource.

Groundwater

Groundwater resource potential

The annual replenishable groundwater resource of the country, as per the year 2006 estimate, is 433 BCM and the net groundwater availability is 399 BCM after allocating 34 BCM for natural discharges during non-monsoon season. The annual groundwater draft for the year 2004 was 231 BCM, out of which 213 BCM is utilized for irrigation and 18 BCM is used for domestic and industrial purposes, as stated above. Overall stage groundwater development is 58 per cent (Chatterjee Rana, 2009). However, 1615 assessment units fall under semi-critical (550), critical (226) and overexploited (839) category, indicating that the groundwater resources in these areas is already being developed, in excess amount than desired for sustainable development. Further areas of the 30 assessment units are completely covered by saline groundwater (Fig.19.1). The stage of groundwater development is high in the states of Delhi, Haryana, Punjab and Rajasthan and Union Territories of Daman & Diu, and Pondicherry, where the overall stage of groundwater development is more than 100 per cent.

Groundwater Development

The percentage of overexploited areas (blocks/talukas/basins) have increased from 4 per cent in 1985 to 15 per cent in 2004 estimation, which is a matter of concern (Planning Commission, 2007). Power subsidy has encouraged greater use of groundwater. The overexploitation has led to depletion of groundwater resources resulting in deeper water levels, widespread and acute scarcity of groundwater. Further this leads to contamination of groundwater due to geogenic factors, resulting in increase in levels of fluoride, arsenic and iron. These problems have serious effect on the rural population where there is no piped water supply or reliable supply of water.

Efficient Water Utilization for Conservation of Groundwater

Improving efficiency of water facilities is to be given priority as the water availability is not to increase dramatically by all means. The National Commission for Integrated Water Resources Development has suggested that the efficiency of the groundwater can be improved from the present level of 65 per cent to 75 per cent. Further waste water treatment, reuse of water, rainwater harvesting and groundwater recharge, watershed development, etc. will also add to the conservation of groundwater thus reducing the gap between demand and supply (Panjiar Umesh Narayan, 2010).

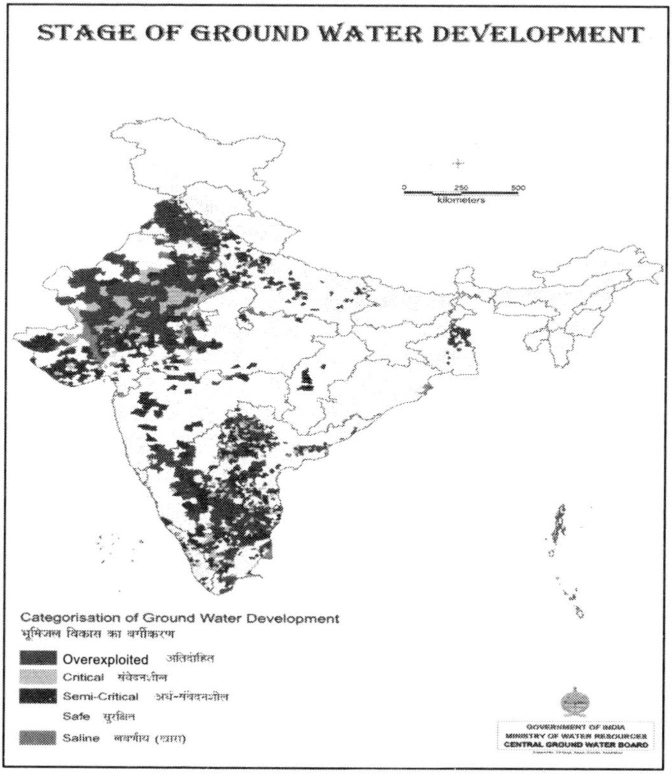

Fig. 19.1: Stage of Groundwater Development in India.

Source: CGWB, 2006.

Groundwater for Poverty Alleviation

The relation between poverty in the country and groundwater development also shows that there is a distinct inverse relationship. The states with higher levels of groundwater development also show lower levels of poverty like Punjab which has only 6.16 per cent of the people below poverty line utilizing 93.34 per cent of its groundwater resources. Similarly, Haryana which has only 8.74 per cent of the people below poverty line utilizing 74.61 per cent of its groundwater resources. On the other hand in Orissa and Bihar where the proportion of people below poverty line is more than 40 per cent, the utilization of groundwater is also very low at about 15.22 per cent and 35.99 per cent respectively (Gupta Archana, *et al.*, 2010). These figures indicate that the groundwater which lies unutilized in these poverty-stricken states can be fruitfully utilized for developing the status of these poor people. This view is also supported by the opinion of Sharma K.D. (2009) who has suggested for improving competitiveness of water-guzzling crops in the groundwater surplus eastern India and less water-requiring commodities in overexploited north west and south India through technology, marketing incentives and disincentive interventions to rationalize the utilization of groundwater in the country. Further, the groundwater sustains the agriculture production in drought years as seen in the year 1987 when the rainfall was 18 per cent below normal and the food grain production declined only by 2 per cent over the previous year's level (Gupta Archana, *et al.*, 2010).

Review of the Existing Groundwater Legislations/Laws

The Indian Easement Act, 1882, is still the relevant Act usually referred to in support of right to groundwater in India. This Act along with its relevant provisions builds upon common law principles establishing a rule of 'absolute ownership' over all there is below the surface of earth of each landowner. Thus, the Indian Easement Act, 1882 links groundwater ownership to the land ownership and this legal position has remained intact since then. However, the right of the owner, in this Act also suggest that it limits an individual's right to (over) exploit groundwater (Planning Commission, 2007). In this context, it will be worth mentioning that, the Hon'ble High Court of Kerala in the famous "Coca Cola case" (*Perumatty Grama Panchayat* Vs. *State of Kerala*) delivered a decision that the State has a right and obligation to restrain use of groundwater if it causes harm to others.

The Indian Constitutional Provisions includes "Water" as a matter under Entry 17 of List II in the Seventh Schedule (i.e. State List). This entry is however subject to the provisions of Entry 56 of List I in the Seventh Schedule, i.e., Union List. Taking into account the word "Water Supply" to include groundwater, will mean that States have jurisdiction to regulate and control groundwater (Planning Commission, 2007).

The Central Ground Water Authority (CGWA) was constituted on 14.01.1997, under sub-section (3) of the Environment (Protection) Act, 1986, in pursuance of Supreme Court Orders on a Public Interest Litigation (PIL), to regulate and control groundwater development and management. Pursuant to the Hon'ble Supreme Court order the CGWA was granted the powers among others to regulate control, manage and develop groundwater in the entire country and issue necessary directions for this purpose.

However, after 13 years of constitution of CGWA, the issue of overexploitation of groundwater remains to be checked, the reasons being many, as seen in many of the urban and semi-urban areas of our country.

Groundwater Problems

The groundwater problem can be broadly divided into two categories: (1) Groundwater resource depletion, and (2) Groundwater quality deterioration. These two problems are known to even the common man and have become an accepted fact that these two problems need immediate attention for sustainable management of water resources.

Groundwater depletion is reduction in the groundwater resource potential. The replenishable groundwater resource is estimated using the groundwater level/peizometric surface and other aquifer parameters for phreatic (water table) and confined aquifers. For simple mathematical computations of storage and flow of groundwater, aquifers are frequently assumed to be homogeneous (no variation through space) and isotropic (no variation with the direction of measurement). Such an idealized aquifers do not exist. However good quantitative approximations are obtained by these assumptions, particularly where average aquifer conditions are employed on a large scale.

The reduction in resource is reflected in the declining water levels and the historical water level records reveal the trend and rate of decline of water level in space and time. The reduction in recharge on one hand due to various factors and steep increase in withdrawal of groundwater on the other hand have resulted in the resource depletion. The depletion of groundwater resource results in deterioration in quality as the process of dilution of the dissolved solids is retarded by reduction in recharge and in some areas this has assumed an alarming situation especially in agricultural lands where there is excessive use of fertilizers and pesticides.

In urban areas, the manifold increase in population and urbanization for their need, coupled with the industrialization has put enormous stress on the groundwater availability. At present most of the cities in India face the problem of water supply as they are not able to cope up with the increasing demand of water supply. The disposal of untreated/partially treated effluents from the industries and urban waste add to this problem resulting in the pollution of the surface water bodies and ultimately reaching the groundwater, remediation of which is not economically viable for a developing country like India.

Growing demands of water in the coming years will trigger acceleration of groundwater development in the coastal tract. But despite large groundwater potentials in the area, groundwater exploitation is fraught with problems arising out of salinity hazards, overdevelopment, rising water table in canal commands and groundwater contamination due to natural or man-made causes. The critical situations as developing in parts of the coastal tract, if unheeded, is likely to cause widespread environmental hazards threatening sustainability of groundwater as a resource.

Problems were also faced during the natural calamities like Super Cyclone in Orissa Coast in 1999, Tsunami affecting the coastal areas as well as the Andaman Island. The primary issue in the Earthquake ravaged Gujarat's Bhuj area was supply of drinking water, which was done on war footing by drilling new tubewells as the existing wells were damaged or polluted.

Management Strategies

The International Conference on Fresh water held at Bonn in 2001 resolved to promote water management strategies at regional, national and local levels. It has emphasized the need to promote water use efficiency, reduce water pollution and conservation of water resources. Water security for all is an achievable goal. There is enough water for everybody, only if we change the way we manage it.

Sustainable Development and Management of Water resources require planning. Planning methodologies require proper understanding of the processes that control occurrence and movement of groundwater, their monitoring in terms of quantity and quality and their interaction with land and surface water resources. The key strategies required to be adopted may be:

1. Long term perspective plan for conservation and augmentation of recharge of groundwater resource,
2. Protection of groundwater quality from deterioration, and
3. Sustainable groundwater development.

All the above three strategies require the community participation for proper implementation and success on the ground.

Water Conservation and Participatory Approach

The groundwater usage is steadily increasing in urban and rural areas due to increase in water demand and as groundwater is a dependable resource. Hence, conservation of groundwater is becoming important day by day. This is possible by adopting the following methods:

(a) By reducing the surface water run-off by construction/ adoption of suitable conservation structures.
(b) By construction of artificial recharge structures wherever technically feasible to augment groundwater recharge.
(c) By reducing the overexploitation of groundwater by enforcement of suitable laws (like Environment Protection (Act) 1986 under which the Central Ground Water

Authority has been constituted for regulation of groundwater development).

It is however all the above processes will require the participation of the public for successful implementation.

An Integrated Approach with Community Participation

An integrated groundwater development and management plan envisaging rational and efficient utilization of groundwater system require a data base, a monitoring network to monitor the ground water regime and their development. The good network of data collection and compilation can involve the stakeholders like the community which uses the groundwater at local levels. The local community can be made aware of the importance of this data by involving local NGOs, who can reach the people. The awareness will initiate involvement of community. The community involvement will improve transparency, which in turn will improve reliability of the data for proper planning.

Community Participation in Basin Management and Conjunctive Use of Surface and Groundwater

Groundwater basin is a hydrogeologic unit containing one large aquifer or several connected and interrelated aquifers. Such a basin may or may not coincide with a physiographic unit. In a valley between mountain ranges the groundwater basin may occupy only the central portion of the stream drainage basin. The concept of a groundwater basin becomes important because of the hydraulic continuity that exists for the contained groundwater resource. In order to ensure continued availability of subsurface water, basin-wise management of groundwater becomes essential. In many cases, surface water basin and groundwater basin are coincident, but topography controls juridiction of surface water basin, whereas the groundwater basin is controlled not only by topography but also by disposition, structure and permeability of rocks and configuration of water table.

The use of surface water in conjunction with groundwater is the concept of conjunctive use, for sustainable development of water and land resources. In the concept of conjunctive use, one of the major concern is decision about the quantity of groundwater use, so that this resource development is sustained by the system. The community needs to be aware of such concepts so that they decide and prioritize the water use in their locality or area.

Community Participation in Managing Groundwater

The groundwater has remained as a community resource in our country from time immemorial. The groundwater was being used as a common resource with common dug wells for the use of all sections of society from historic time, thus binding the community for cooperative use of natural resource. This system is continuing even today in most of the villages, though in some prospering states, the dug wells are replaced by tube wells or bore wells due to depletion of the groundwater table. However, this is not true in urban areas and also with the changing time the rural areas will also start changing. This development is due to affordability of an individual or a farmer. But this is a matter of concern from the resource point of view, as it only leads to depletion of the groundwater resource, with increase in withdrawal of groundwater, with manifold growth in population, urbanization and industrialization.

Involvement of Community

The involvement of community through participatory approach is essential. The success stories of Hiware Bazar village in Ahmed Nagar district of Maharashtra and Neemi village of Jaipur district of Rajasthan have shown that the key to success was involvement of the stakeholders, the community itself. It was observed when the community is empowered they prioritize the needs, take appropriate decisions and manage the water resources without any conflicts, as it gives them a sense of accountability towards each other. Further when they are aware about the pros and cons of the depletion of groundwater they are forced to think for sustainable development.

Involvement of NGOs

The NGOs who are working at the grassroot level are the right people to convey the message of conservation of groundwater and make them understand the need for sustainable development to avoid depletion of the dwindling resource. The NGOs can reach the people, in the present system of governance.

Awareness

The awareness about the status of groundwater at local level has to be made to the public through various media and also through involvement of various organizations working in that area, Panchayati Raj Institutions and Resident Welfare Associations in the urban areas. The process of these awareness programmes can be monitored using various indices by the funding agencies and deficiency if any can be rectified from time to time for appropriate progress in conservation of fresh groundwater resources, thus creating a sustainable community which is aware of the status of groundwater of its area.

Conclusion and Recommendations

The general decline in groundwater resources and deterioration in quality of groundwater all-over the country are indications of mismanagement of groundwater resources. The studies have revealed that the rainfall contributes 67 per cent of the country's annual replenishable groundwater resource. The drought years are generally sustained by the groundwater irrigation. The efficiency of groundwater utilization can be improved in the areas where there is overdevelopment of groundwater to conserve the fresh water. However, the development of groundwater can be encouraged in areas, especially eastern and North-eastern parts of the country, where the groundwater development is low, for prosperity of these areas by augmenting the agricultural production in these areas through groundwater irrigation, by improving the reliability of power supply.

There is an urgent need for creation of a community which is empowered and aware of importance of conservation and augmentation of groundwater resource. The involvement of communities in management of groundwater will lead to sustainability of this invisible resource. The government policy should aim at encouraging the involvement of communities for development and management of groundwater as the regulation of groundwater will be an impossible task for the Governmental agencies, without involving the communities as seen through experience. Good practices are to be promoted with incentives so that such practices spread to other parts of the country. Practicing makes the difference on the ground than the laws which remain in the papers without implementation.

REFERENCES

Anandha Kumar, K.J., Vijay Kumar, Maanjhu, S.K and Das, P. (2007). Groundwater Management Options—A case study of Western Yamuna Canal Command, *Haryana—Water Digest,* Vol. I, Issue 6, pp. 30-35.

Chatterjee, Rana and Purohit Raja Ram (2009). "Estimation of replenishable groundwater resources of India and their status of utilization." *Current Science,* Vol. 96(12): 1581-1591.

Das, S. and Anandha Kumar, K.J. (2005). "Management of Ground water in Coastal Orissa", *International Journal of Ecology and Environmental Science,* Vol. 31(3) 3, Special Issue, pp. 285-297.

Gupta Archana and Shukla Akanksha (2010). "Fighting Poverty through Water Harvesting". *Yojana,* ISSN 097-8400, pp. 45-47, July, 2010.

Ministry of Water Resources (1999). Integrated Water Resource Development—A Plan for Action—Report of National Commission for Integrated Water Resources Development—Volume-I.

Panjiar, Umesh Narayan (2010). "Efficient Water Managemnet: Challeng and Initiatives." *Yojana,* pp. 5-8, July, 2010.

Planning Commission, (2007). Report of the Expert Group on Ground Water Management and Ownership.

Sharma, K.D., (2009). "Groundwater management for food security." *Current Science,* Vol. 96, No. 11, June, 2009.

20

Trends in Corporate Social Responsibility

From Charity to Sustainability

K. Soundararajan

Introduction

Today India has emerged as a global leader in information technology, business process outsourcing, telecommunications and pharmaceuticals. Literacy levels have gone up and better health care has improved the quality of life. However, India's impressive growth also comes with the burden of a rising population.

Parallel to economic growth, unemployment rates have risen and there is skill shortage for jobs in modern industries. Low agricultural productivity has compounded farmers' distress in many parts of the country. Also, one of the main challenges before India is to provide adequate, affordable fossil and renewable energy for development of all, at the same time addressing social inequities and income disparities in this regard.

As corporates have been regarded as institutions that cater to the market demand by providing products and services, and have the onus for creating wealth and jobs, their market position has traditionally been a function of financial performance and profitability. However, over the past few years, as a consequence of rising globalisation and pressing ecological issues, the perception of the role of corporates in the broader societal context within which it operates, has been altered. Stakeholders (employees, community, suppliers and shareholders) today are redefining the role of corporates taking into account the corporates' broader responsibility towards society and environment, beyond economic performance, and are evaluating whether they are conducting their role in an ethical and socially responsible manner. As a result of this shift (from purely economic to 'economic with an added social dimension'), many forums, institutions and corporates are endorsing the term Corporate Social Responsibility (CSR). The concept of CSR in India is not new, the term may be. The process though acclaimed recently, has been followed since ancient times albeit informally.

Philosophers like Kautilya from India and pre-Christian era philosophers in the West preached and promoted ethical principles while doing business. The concept of helping the poor and disadvantaged was cited in much of the ancient literature. The idea was also supported by several religions where it has been intertwined with religious laws. "Zakaat", followed by Muslims, is donation from one's earnings which is specifically given to the poor and disadvantaged. Similarly, Hindus follow the principle of "Dhramada" and Sikhs the "Daashaant".

The term CSR itself came in to common use in the early 1970s although it was seldom abbreviated. By late 1990s, the concept was fully recognized; people and institutions across all sections of society started supporting it. This can be corroborated by the fact that while in 1977 less than half of the Fortune 500 firms even mentioned Corporate Social Responsibility (CSR) in their annual reports, by the end of 1990, approximately 90 per cent Fortune 500 firms embraced CSR as an essential element in their organizational goals, and actively promoted their CSR activities in annual reports (Boli and Hartsuiker, 2001).

According to World Business Council for Sustainable Development "Corporate Social Responsibility is the continuing commitment by business to behave ethically and contribute to economic development while improving the quality of life of the workforce and their families as well as of the local community and society at large". CSR is a link to sustainable development. Businesses feel that there is a need to integrate social, economic and environmental impact in their operation.

Definition #3: Archie Carroll in 1991 describes CSR as a multilayered concept that can be differentiated into four interrelated aspects—economic, legal, ethical and philanthropic responsibilities.

Changing Public Expectations of Business

Globally companies are expected to do more than merely providing jobs and contribute to the economy through taxes and employment. Consumers and society in general expect more from the companies whose products they buy. This is coherent with believing the idea that whatever profile is generated is because of society, and hence mandates contributing a part of business to the less privileged.

Trends in CSR: From Charity to Sustainability

An insight into the history of CSR reveals that till 1990s it was solely dominated by the idea of philanthropy. Considering CSR as an act of philanthropy, businesses often restricted themselves to one time financial grant and did not commit their resources for such projects. Moreover, businesses never kept the stakeholder in mind while planning for such initiatives, thereby reducing the efficacy and efficiency of CSR initiatives. However, over the last few years, the concept of CSR has been changing. There has been an apparent transition from giving as an obligation or charity to giving as a strategy or responsibility. Review of the case studies and work done on CSR by companies in India suggests that the CSR is slowly moving away from charity and dependence and starting to build on empowerment and partnership.

Today, Corporate Social Responsibility (CSR) has been understood in terms of accountability where corporates are feeling that they are responsible for the impact their actions have on several stakeholders. They feel that the basic motive of CSR today is to increase the company's overall impact on the society and stakeholders.

The following are key focus areas being incorporated into business practices:

- **Partnership with NGOs and Government:** There has been an increase in the number of corporates partnering with NGOs and the Government to ensure successful execution of initiatives.
- **Community Development:** Most large companies either have their own foundations or contribute to other initiatives that directly support the community upliftment, notably in health, education, and agriculture.
- **Environmental Management:** Environmental policies and programmes are now standard, and many companies have implemented the ISO 14001 system throughout their businesses.

Workplace

Growing out of a long-standing commitment to training and safety is a more recent emphasis on knowledge and employee well-being.

Sustainability Reporting (SR) is also gaining prominence and recognition as a value added tool for displaying a corporate's commitment towards transparency and accountability towards its stakeholders. Sustainability is the new lingua franca in the market place as Indian corporates start thinking beyond profits about the welfare of the planet.

Though there have been evidences that record a paradigm shift from charity to a long-term strategy, yet the concept still is believed to be strongly linked to philanthropy. There is a need to bring about an attitudinal change in people about the concept. Several corporates today have specific departments to operationalise CSR. There are either foundations or trusts or a separate department within an organization that looks into the implementation of practices. Sustainable development, like building a successful business, requires taking the long-term view. The KPMG International Survey of Corporate Responsibility Reporting, 2005 showed that voluntary reporting on sustainability is on the increase across all the countries. Sustainability Reporting is emerging as a key vehicle to implement CSR and measure its progress in organizations.

Businesses have begun to respond. Global industrial giants, from Pepsico to Walmart and IBM, all recognize that global warming is the new reality in the marketplace. And it is not just environmental woes; as business leaders come face-to-face with a slew of complex social and governance challenges that present both risks and opportunities to performance and reputation they are viewing business operations through new lens—sustainability. "A sustainable organization runs a profitable operation while continually contributing towards the betterment of all stakeholders. If any stakeholder whether the environment or your employee shows deterioration, you are no longer sustainable."

Business Does Well by Doing Business

The purpose of any business that seeks to be sustainable has to be more than generating short-term shareholder value. Simply by adding the word long-term to share value, they embrace everything necessary for the survival and success of the company. This includes building trust among communities and maintaining a healthy environment to do business. All of these benefits are created in the normal course of responding to the market signals.

More companies across markets and industries provide many positive examples of actions to integrate climate change and natural resource management in their governance practices and strategic planning.

Progressive businesses are widening their horizons as they look for opportunities and are gaining competitive advantage by responding to societal signals as well as market signals. For example, a company that acts to promote human welfare is helping to build its future market and workforce. A company that acts to protect the environment is helping to preserve the natural capital upon which its operations depend and possibly gaining early-mover advantage in the more highly-regulated markets of the future. Such companies realize that ignoring these issues means—at best—missing early indicators of future profitability and—at worst—risking the company's market valuation and reputation.

In responding to societal signals, companies are now looking more widely and imaginatively at the fundamental purpose of providing continuous improvement in goods and services to increasing numbers of customers at prices they can afford. They are improving their products in ways that serve societal ends, and they are increasing customer numbers by reaching out to new markets in developing and emerging economics. In this, the role of business, shareholder value is seen as the measure of success in fulfilling the more fundamental purpose of providing improving goods and services that today's and tomorrow's consumers want. As a result of responding to societal signals, many companies are now active in areas often considered the domain of governments; providing basic goods such as health, education, and pensions, advocating a framework to tackle climate change, or addressing poverty through corporate strategy.

Measuring CSR

The development of corporate governance ratings and surveys has become a measure of companies image. For example, Indian market has developed a yardstick in the form of 'Governance and Value Creation Rating (GVC)'. This was developed by CRISIL. This is significantly different from the 'Traditional Corporate Governance Ratings (TCGs)', which is usually available in Business—Moving Towards BOP (Bottom of the Pyramid)

The Bottom of the Pyramid debate has led to new perspectives on the opportunity. There is a focus on the Middle of the Pyramid, or the aspiring middle class. A recent study by the *Economist* concluded that half the world can be classified as the emerging middle class; defined as a population living on $2-13 at 2005 Purchasing Power Parity Prices. They have discretionary income and spend on education, health, energy, transportation, and personal care. This market by some estimates includes 2.6 billion people in 2005 and is rising fast. Asia alone is expected to have approximately 60 per cent of the global middle class. Many firms now classify the Bottom of the Pyramid market as "emerging consumer markets" or just emerging markets. Obviously, this is a more emotionally neutral term. More important, this view of Bottom of the Pyramid markets—the emerging consumers—shows the respect and the commitment of managers in large firms who have spent time and resources to understand this opportunity.

Sustainability will emerge as a source of innovations at the Bottom of the Pyramid: The planet is under stress—be it access to water, deforestation, pollution, or greenhouse gas emissions. This is a familiar story. What if we added an additional 3 to 4 billion people as micro producers and consumers to the current 2 billion? Sustainability will become a major impediment. Nobody can deny that the poor have a right to a better standard of living—be it access to safe water, better food and sanitation, and better education. The pressure to serve an additional 3 to 4 billion and at the same time protect the environment will focus attention on sustainability as never before.

Sustainable Infrastructure: A Necessity, Not an Alternative

Economic slowdown offers governments and businesses the opportunity to increase their spending on sustainable infrastructure as a stimulus to avoid severe recession at the same tackling environmental and social conflict with infrastructure. The opportunity is huge. For instance, clean energy infrastructure requires annual investments of a minimum of $515 billion until 2030 as suggested by the World Economic Forum. Looking beyond energy infrastructure, other few trillion-dollar investments and social policy frameworks are required to develop sustainable infrastructure.

Emerging Business Cases in India

Sl. No.	*Company Name*	*Thematic Areas*
1.	Infosys Technologies Ltd.	Education, Health, Environment
2.	Ambuja Cements Ltd.	Poverty Reduction, Environment
3.	CoCa-Cola India Ltd.	Environment, Educational Livelihood
4.	Dr. Reddy's Laboratoris	Micro Finance, Health, Environment, Social enterprisers
5.	Reliance Industries Ltd.	Education Environment, Child welfare and infrastructure development
6.	Tata group	Wants all their companies to cut emissions drastically

(*Contd.*)

Sl. No.	*Company Name*	*Thematic Areas*
7.	ITC	3000 hectares of the afforestations project is registered under the United Nations Framework convention on climate change, which means it will be eligible for carbon credits.
8.	IOC	Beyond Petroleum every oil company in its heart, knows that just crude is the longer a sustainable business. It started working on a hydrogen CNG fuel.
9.	WIPRO	Has always lean proactive on issues like efficient use of water and energy, decided to make ecological sustainability a part of its core values.
10	HSBC	Environment initiatives are targeted towards nature, habitat and biodiversity conservation, water harvesting and climate change initiatives

Conclusion

Corporate Social Responsibility is not a fad or a passing trend, it is a business imperative that many Indian companies are either beginning to think about or are engaging with in one way or another. While some of these initiatives may be labeled as corporate citizenship by some organizations, their basic message and purpose is the same. A successfully implemented CSR strategy calls for aligning these initiatives with business objectives and corporate values thereby integrating corporate responsibility across the business functions and enhancing business reputation.

The challenge for us is to apply fundamental business principles to make CSR sharper, smarter, and focused on what really matters.

REFERENCES

CII-ITC, "Winning strategies for a sustainable world", 4th Sustainability summit 2009, Report.

Corporate Social Responsibility Towards a Sustaihable Future-KPMG in India Report.

GTZ-Gesellschaft for Technische Zusammenarbeit.

Outlook Business—Green companies—May 2010.

Sustaihasle Development and Environmental issues"—CYP Module-13.

The Economic Times"—Nov. 22, 2009.

The Fortune at the Bottom of the Pyramid"—C.K. Prahald.

www.worldvision.in/corp.

21

Meltwater Characteristics of Raktvarn Nala of Gangotri Glacier, Garhwal Himalaya, India

A.L. Ramanathan, Virendra Bahadur Singh and *P.G. Jose*

Introduction

Himalaya is one of the youngest, tallest and the steepest mountain system on the Earth. Major rivers in the subcontinent lies in the upper part of the Himalaya. It is covered by glacier and snow (Ahmad *et al.*, 1999). Glacierised areas present an ideal environment in which to study water-rock interaction, since chemical weathering rates are high and anthropogenic impacts are often minimal (Brown, 2002) for example, the Himalaya constitutes one of the most important glacier systems of the world, contributing significantly to the continental solute budget. In India, glaciers are restricted to within the latitude 27°-N to 36°-N and longitude 72°-E to 96°-E (Glacier Atlas of India, 2008). The inventory compiled by the Geological Survey of India mentions over 9000 glaciers—9,575 to be exact—in the Indian administered part of the Himalayas comprising the territories of Jammu & Kashmir, Himachal Pradesh, Uttarakhand, Sikkim and Arunachal Pradesh. Total glacier cover within these states is less than 40,000 km^2 with an approximate volume of ice of the order about 2,000 Km^3 (Sangewar & Bajpai, 2009). These glaciers, being sources for the principal river systems originating in the Himalayas, form an important source of water and maintain water supply in these rivers throughout the hot and dry summer months (Bahadur, 1988).

Various studies of glaciers specially in Himalaya by Ahmad and Hasnain (1999), Kumar *et al.* (2008) and in Alpine region by

Rainwater and Guy (1961), Beherens *et al.* (1971), Lorrain and Souchez (1972), Collins (1978), Tranter and Raiswell (1991), have suggested that principal flow component in water passing through subglacial channels had good exposure to subglacial environment. These studies indicate high rates of chemical weathering in glaciated catchments resulting from rapid dissolution of freshly ground glacial rock flour. Hydrochemistry of glacier evolved from precipitation (rainfall and snow) and get enriched during the traverse through sub-glacial and ice rock interface (Trudegill, 1986).

Area of Study

Gangotri glacier (79°17′18′′ to 79°4′55′′ E and 30°43′10′′-30°55′50′′ N) is the largest glacier in the Himalayas and is located in the Uttarkashi district of Uttarakhand State (Kumar *et al.*, 2008). Although it is commonly known as Gangotri glacier, but actually it is the Gangotri glacier system consisting of a cluster of many large and small glaciers (Fig. 21.1). Gangotri glacier system consists of three major glacier tributaries, known as, Chaturangi glacier (length 22.45 km; area 67.70 km^2), Kirti glacier (length 11.05 km; area 33.14 km^2) and Raktvarn glacier (length 15.90 km; area 55.30 km^2) with main Gangotri glacier (length 30.20 km; area 86.32 km^2) as trunk part of the system (Singh *et al.*, 2005).

Materials and Methods

Melt water samples draining from Raktvarn Nala of Gangotri Glacier were collected from August-September 2007. Electrical Conductivity (EC) was measured by using Pentex EC meter. pH of all samples were measured by Consort microcomputer (P-307) ion meter. Suspended sediments were separated from the water samples in the laboratory by using 0.45 micro millipore membrane filters of 47 mm diameter. Vacuum pump was used to accelerate the filtration. Bicarbonate was determined by following potentiometric titration methods (APHA, 2005), Chloride ion concentration was measured by the mercury (II) thiocynate method, Sulphate concentration was measured by turbidimetric method (APHA 2005), the dissolved silica was determined by molybdosilicate method (APHA 2005), Nitrate concentration in water samples was determined by brucine-sulphanilic acid method

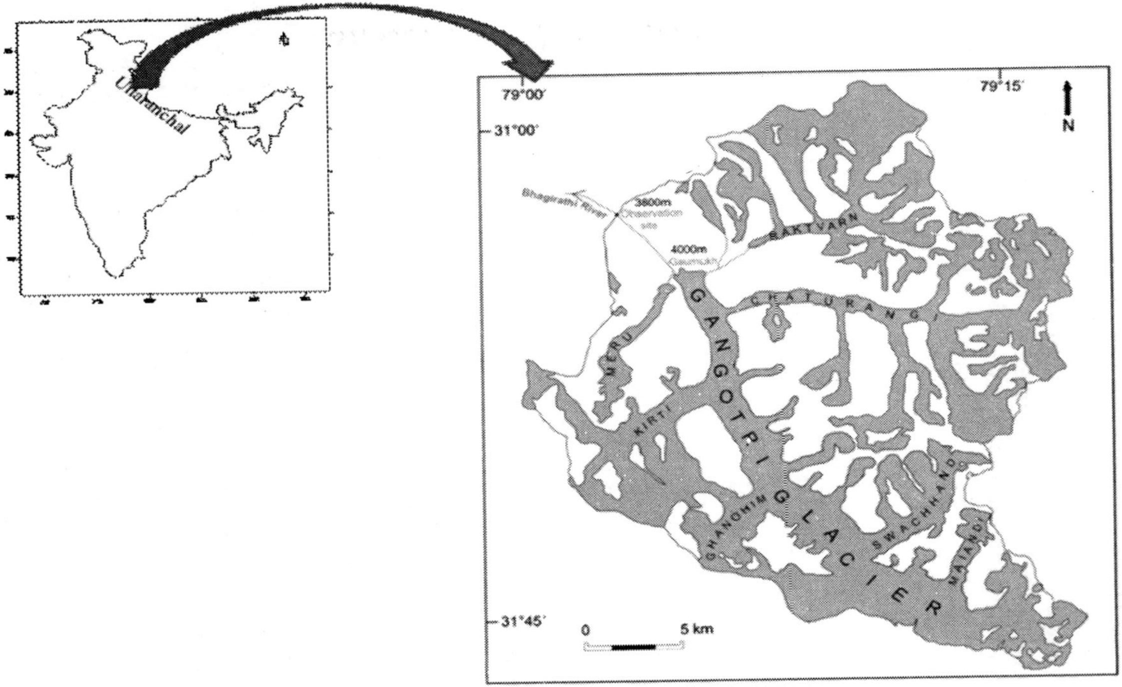

Fig. 21.1: Map showing Location of the Study Area with the Gangotri Glacier System.
Source: Singh, *et al.*, 2005.

and the content of major cations Na^+, K^+, Ca^{2+}, Mg^{2+} was determined on GBC 906 atomic spectrophotometer (AAS).

Results and Discussion

The chemical characteristics of melt water of Raktvarn Nala of Gangotri Glacier is shown in Table 21.1. The average EC and pH of melt water of Raktvarn nala of Gangotri glacier was recorded as 214.4 μs/cm (257.8-174.4 μs/cm) and 5.8 (5.5-6.1) respectively. The pH of Himalayan glacier melt waters is significantly higher than pH of the melt water of the North Cascade glaciers (Reynold and Johnson, 1972). The average concentration range of major cations was found to be Ca^{2+} 743.4 μeq/l (614.9-850.0 μeq/l) and Mg^{2+} 582.8 μeq/l (550.0621.8- μeq/l) followed by K^+ 117.5 μeq/l (87.2-160.4 μeq/l) and Na^+ 79.9 μeq/l (59.5-100.4 μeq/l) respectively. The average chemical composition of melt water showed major percentage of Ca^{2+} (49.0 % of total cations) followed by Mg^{+2} (38.4% of total cations), K^+ (7.7% of total cations) and Na^+ (5.2% of total cations). The average concentration of major anions was SO_4^{2-} 1701.7 μeq/l (1493.7-1905.2 μeq/l) followed by HCO_3^- 143.2 μeq/l (89.8-201.9 μeq/l), Cl^- 34.9 μeq/l (5.5-72.2 μeq/l) and NO_3^- 4.6 μeq/l (2.9-6.6 μeq/l) respectively. The average chemical composition of melt water showed major percentage of SO_4^{2-} (90.6% of total anions) followed by HCO_3^- (7.6% of total anions), Cl^- (1.8% of total anions) and NO_3 (0.24% of total

Table 21.1: Maximam, Minimam, Average and Standard Deviation Value of Different Parameter of Melt Water of Raktvarn Nala of Gangotri Glacier in August-September 2007

Parameter	*Maximum*	*Minimam*	*Average*	*SD*
EC	257.8	174.4	214.4	37.8
pH	6.1	5.5	5.8	0.3
Ca^{2+}	850.0	614.9	743.4	102.0
Mg^{2+}	621.8	550.0	582.8	35.1
K^+	160.4	87.2	117.5	32.1
Na^+	100.4	59.5	79.9	17.7
$SO4^{2-}$	1905.2	1493.7	1701.7	188.8
HCO^{3-}	201.9	89.8	143.2	46.5
Cl^-	72.2	5.5	34.9	29.6
NO^{3-}	6.6	2.9	4.6	1.7
H_4SiO_4	119.0	66.5	94.2	24.6
TSM	924.0	648.0	799.0	126.6

Unit: Dissolved ions in μeq/l, H_4SiO_4 in μmole/l, TSM in mg/l.

anions). The average value of dissolved silica was found to be 94.2 μmole/l (66.5-119.0 μmole/l). The average value of Total Suspended Matter (TSM) was found to be 799.0 mg/l. (648.0-924.0 mg/l).The major cations contribution was found to be nearly 44.6 per cent of the Total Dissolved Solid (TDS) while major anion contribution was nearly 55.4 per cent of the Total Dissolved Solids (TDS).

The chemical composition of glacier run-off indicates the nature of chemical weathering taking place beneath the glacier (Lorrains and Souchez, 1972; Raiswell, 1984). Catchments of glaciers undergo more intense chemical weathering than the catchments which do not have glaciers (Reynolds and Johnson, 1972; Collins, 1983). The solute content present in melt water is derived from two sources, lithospheric and atmospheric. The release of solute to glacier melt waters is governed by the interaction between the rock water interface, types of weathering reactions which are taking place, mineralogy of the bedrock, flow rate of water and the composition of atmospheric deposition in the catchment of glacier.

The chemical composition of meltwater with reference to major ions can be explained on the basis of weathering of various rock forming minerals. In the glacier environment, the proton H^+ used for chemical weathering of carbonates, sulphides and aluminosilicates are mainly supplied from the equilibration of meltwater with atmospheric CO_2 and/or by oxidative weathering of sulphides (Raiswell, 1984). Hence Ca^{+2}, Mg^{+2} and HCO_3^- in aquatic environment can be derived from the weathering of carbonate and/or silicate by the reaction.

$$CO_2 + H_2O \rightarrow H_2CO_3$$
$$H_2CO_3 \rightarrow H^+ + HCO_3^-$$
$$4\ FeS_2 + 15\ O_2 + 8\ H_2O \rightarrow 2\ Fe_2O_3 + 8\ SO_4^{2-} + 16\ H^+$$

Carbonate minerals notably Calcite and Dolomite reacts with carbonic acid and released HCO_3^- and cations to the solution by the reactions.

$$CaCO_3 + H_2CO_3 \rightarrow Ca^{2+} + 2\ HCO_3^-$$
$$Ca\ Mg\ (CO_3)_2 + 2\ H_2CO_3 \rightarrow Ca^{2+} + Mg^{2+} + 4\ HCO_3^-$$
$$(Ca, Mg, Na, K)\ Silicates + H_2CO_3 \rightarrow Ca^{2+} + Mg^{+2} + Na^+ + K^+ + H_4SiO_4 + HCO_3^- + Clay\ minerals$$

Hence two major anions HCO_3^- and SO_4^{2-} in melt water of glacier are mainly derived from the dissolution of atmospheric CO_2 in H_2O and the oxidation of sulphides or dissolution of gypsum. The dissolution of carbonate rocks proceeds more rapidly than silicate breakdown and is the likely mechanism of solute acquisition. The relative proportion of the various ions in solution is depending on their relative abundance in the host rock and their solubility.

The Ca+Mg Vs TZ^+ plot in Fig. 21.2(a) shows that the concentration of Ca^{2+} + Mg^{+2} mostly fall along 1:1 line with an average equivalent ratio of 0.87. It is indicated that the carbonate weathering could be the major source of dissolved ions in the melt water draining from Raktvarn Nala of Gangotri Glacier. The Na + K Vs TZ^+ plot in Figure 21.2(b) shows that there is a small contribution from silicate weathering. The relative high contribution of Ca+Mg to the total cations, the high Ca+Mg/Na+K ratio (6.71) and the low (Na+K)/TZ^+ ratio (0.13) also confirms that carbonate weathering could be major source of dissolved ion in the melt water of Raktvarn nala of Gangotri glacier.

The variation of weight ratio of Na/(Na +Ca) as a function of total dissolve solids (TDS) has been plotted to decipher the mechanism controlling water chemistry (Gibbs, 1970). This plot (Fig. 21.3) places the Raktvarn nala of Gangotri glacier in the region of rock dominance which is consistent with the observation that the dissolution of various rock forming mineral in the water is a primary factor controlling the chemistry of meltwater draining from Raktvarn nala of Gangotri glacier.

Low concentration of NO_3^- at the sampling site was reported which indicated that nitrate is derived exclusively from the atmosphere associated with acidic nitrate aerosols (Tranten *et al.*, 1993).

The thermodynamic stability relationships of water chemistry in the Na^+, K^+, Ca^{2+} and Mg^{2+} silicate system is demonstrated in Fig. 21.4 (after Garrels and Christ, 1965). The plot of Na, K, Ca and Mg silicate system for the melt water draining Raktvarn nala of Gangotri glacier is in the range of the stability field of Kaolinite, which implies that the chemistry of the water favours Kaolinite formation. This is also needed for X-ray mineralogical study on suspended sediments. We observe low concentration of dissolve silica in the melt water draining from Raktvarn nala of Gangotri Glacier, it may be attributed to high resistance of silica mineral to

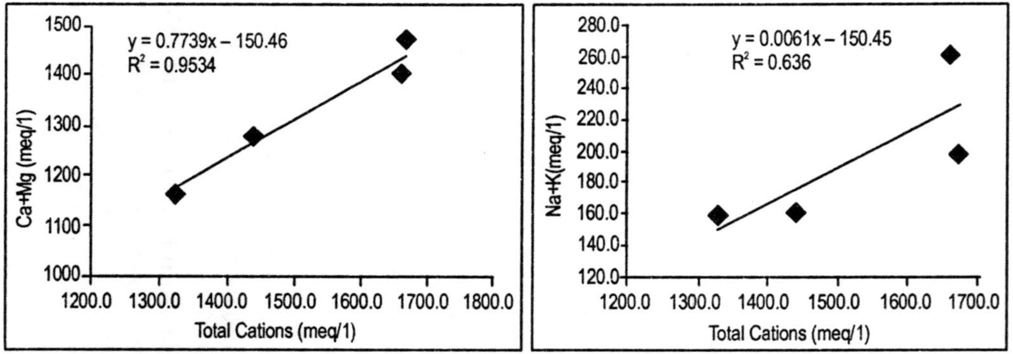

Fig. 21.2: (a) Scatter Diagram Between Ca^{2+}+Mg^{2+} Vs. Total Cations (TZ^+)
(b) Scatter Diagram Between Na^++K^+ Vs. Total cations (TZ^+)

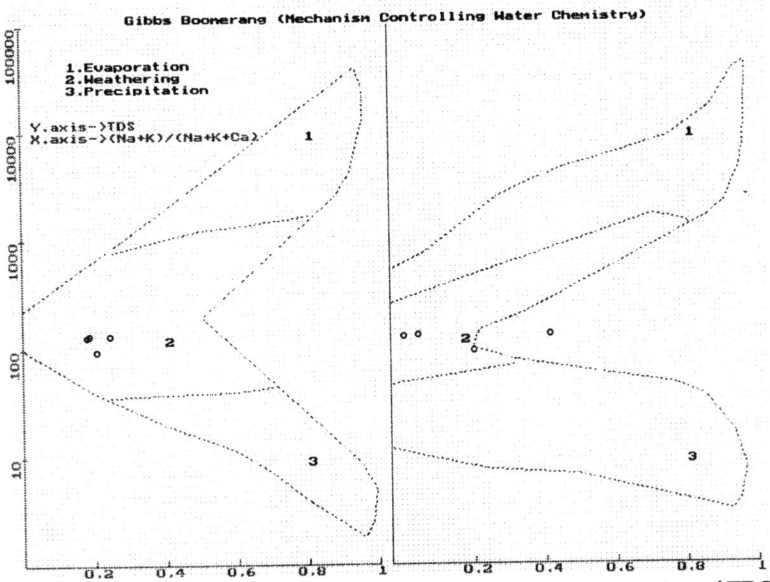

Fig. 21.3: Variation of Weight Ratio ($Na^+/Na + Ca^{2+}$) as a Function of TDS (After Gibbs 1970).

weathering and also consumption of H_4SiO_4 in the formation of secondry mineral (kaolinite).

A parametric correlation analysis for average values of all measured parameters is given in Table 21.2. EC is highly correlated with H_4SiO_4 (r^2=0.788), NO_3 (r^2=0.966), SO_4 (r^2=0.794), Ca (r^2=0.788), Mg (r^2=0.988), Na (r^2=0.701), and TDS (r^2=0.887). H_4SiO_4 is highly correlated with NO_3 (r^2=0.833), SO_4 (r^2=0.875), Mg (r^2=0.895), Na (r^2=0.819), K (r^2=0.843) and TDS (r^2=0.901), NO_3 is highly correlated with SO_4 (r^2=0.839), Ca (r^2=0.913), Mg (r^2=0.992), Na (r^2=0.761) and TDS (r^2=0.926). Cl is highly correlated with K (r^2=0.890) and TSM (r^2=0.811). SO_4 is highly correlated with Ca (r^2=0.887), Mg (r^2=0.856) Na (r^2=0.99), K (r^2=0.814) and TDS (r^2=0.982), Mg is highly correlated with Na (r^2=0.777) and TDS (r^2=0.937). Na is highly correlated with K (r^2=0.831) and TDS (r^2=0.948).

Ca^{2+} source showed good correlation with SO_4^{2-}(r^2= 0.887), indicating that source of calcium in meltwater of Raktvarn nala of Gangotri glacier is gypsum. EC and TDS (r^2=0.887) showed a good correlation in meltwater of Raktvarn nala of Gangorti glacier because conductivity increases as the concentration of all dissolved constituents.

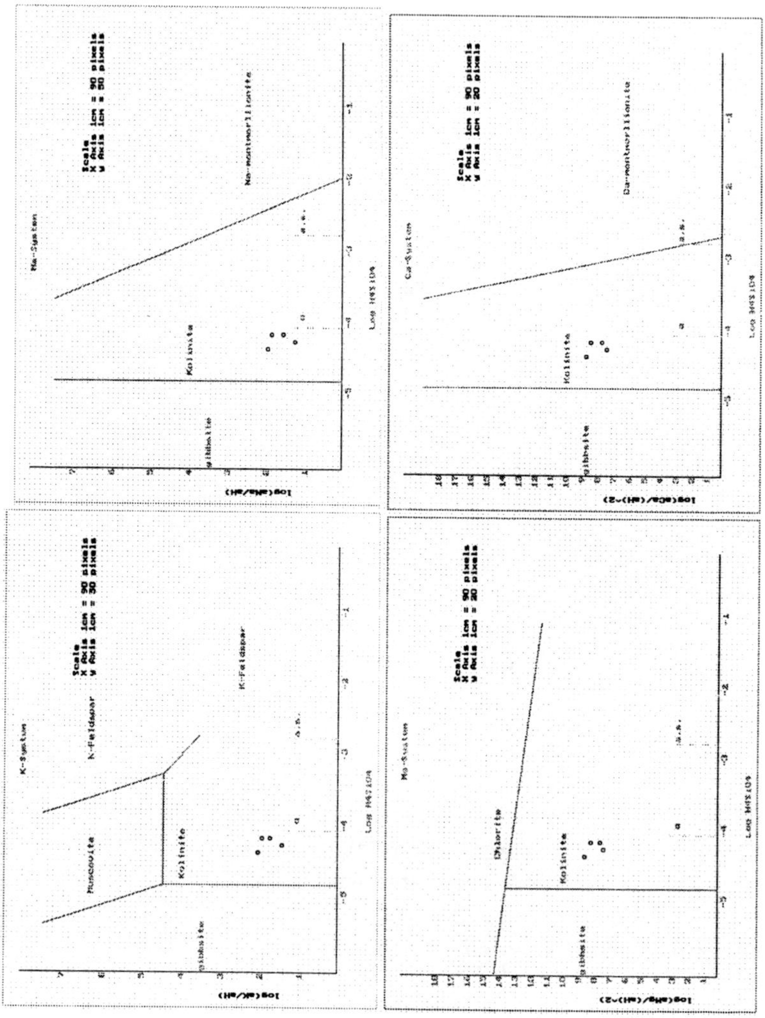

Fig. 21.4: Stability Diagram for Na, K, Ca, Mg Silicate System.

Table 21.2: Parametric Correlation Matrix of Measured Parameters of Meltwater of Raktvarn Nala of Gangotri Glacier

	EC	*pH*	*H_4SiO_4*	*NO_3*	*Cl*	*SO_4*	*HCO_3*	*Ca*	*Mg*	*Na*	*K*	*TSM*	*TDS*
EC	1.000												
pH	-0.128	1.000											
H_4SiO_4	0.910	-0.392	1.000										
NO_3	0.966	0.121	0.833	1.000									
Cl	0.256	-0.800	0.626	0.099	1.000								
SO_4	0.794	0.012	0.875	0.839	0.461	1.000							
HCO_3	0.355	0.553	-0.065	0.440	0.790	0.076	1.000						
Ca	0.788	0.408	0.680	0.913	0.002	0.887	0.347	1.000					
Mg	0.988	-0.005	0.895	0.992	0.215	0.856	0.355	0.875	1.000				
Na	0.701	0.037	0.819	0.761	0.487	0.990	0.175	0.857	0.777	1.000			
K	0.552	-0.502	0.843	0.479	0.890	0.814	0.574	0.454	0.562	0.831	1.000		
TSM	0.204	-0.997	0.459	0.044	0.811	0.053	0.521	0.340	0.082	0.021	0.541	1.000	
TDS	0.887	0.033	0.901	0.926	0.366	0.982	0.092	0.924	0.937	0.948	0.737	0.038	1.000

Conclusion

Melt water of Raktvarn nala of Gangotri glacier is acidic in nature. SO_4^{2-} is dominant anion followed by HCO_3^- and Ca^{2+} is dominant cation followed by Mg^{2+}. The relative high contribution of $(Ca^{2+}+Mg^{2+})$ to the total cations, the high $(Ca^{2+}+Mg^{2+})/(Na^++K^+)$ ratio (6.71) and the low $(Na^++K^+)/TZ^+$ ratio (0.13) indicate that carbonate weathering is the major source of solute in the melt water of Raktvarn nala of Gangotri glacier. Mg^{2+}, Ca^{2+} and HCO_3^- is derived from the weathering of carbonate and/or silicates and SO_4^{2-} by the oxidation of sulphides or dissolution of gypsum. Low concentration of nitrate at the sampling site was reported which indicates that nitrate is derived exclusively from the atmosphere associated with the acidic nitrate aerosols. The variation of weight ratio of $Na^+/(Na^++Ca^{2+})$ as a function of Total Dissolved Solid (TDS) has been plotted to decipher the mechanism controlling the water chemistry. This plot places the Raktvarn nala of Gangotri glacier in the region of rock dominance which is consistent with the observation that the dissolution of various rock forming mineral in the water is a primary factor controlling the chemistry of melt water draining from Raktvarn nala Gangotri glacier. The stability diagram relationship demonstrates that chemical composition of the melt water draining form of Raktvarn nala of Gangotri glacier is in the region of Kaolinite.

REFERENCES

Ahmad, S. and Hasnain, S.I. (1999). "Hydro-Geochemical characteristics of meltwater draining from Bagni glacier, Dhauliganga basin, Garhwal Himalayas, Uttar Pradesh." *J. Appl Hydrol* XII (2 and 3), 53-60.

Ahmad, S. and Hasnain, S.I. (2001). "Chemical characteristics of stream draining from Dudu glacier: An alpine meltwater stream in Ganga headwater, Garhwal Himalaya." *J. of China University of Geosciences*, Vol. 12(1), 75-83.

APHA (2005). "Standard methods for examination of water and wastewater", 21st edn. *American Public Health Association, Washington*, p. 6710.

Bahadur, J. (1988). Himalayan water from snow and glaciers. In First National water conference proceedings, Vol. II. Ministry of Water Resources, Government of India, pp. 59-65.

Behrens, H., Bergmann, H., Moser, H., Rauert, W., Stichler, W., Eisner, H. and Pessel, K. (1971). "Study of the discharge of alpine glaciers by means of environmental isotopes and dye tracers." *Zeitscherkd Glazialgeo* 7, pp. 79-102.

Berner, E.K. and Berner, R.A. (1987). *The global water cycle: geochemistry and environment.* Prentice Hall, Englewood Cliffs, p. 397.

Brown G.H., Sharp M., Tranter M. (1996). "Subglacial Chemical Erosion: Seasonal Variation in Solute Provenance, Haut glacier d Arolla, Valais, Switzerland." *Annals of Glaciology*, 22, pp. 25-31.

Brown, G.H. (2002). "Glacier meltwater hydrochemistry." *Applied Geochemistry*, 17, pp. 855-883(29).

Chidambaram S., Ramanathan, A.L. (2000). Ph.D Thesis, Annamalai University, Tamil Nadu.

Collins, D.N. (1978). "Hydrology of an alpine glacier as indicated by the chemical composition of meltwater." *Zeitschrift fur Gletscherked Glazialgeologieed*, 13:219-238.

Collins, D.N. (1983). "Solute yield from a glacierized high mountain basin." In: *Dissolved Loads of Rivers and Surface Water Quantity/Quality Relationships* (ed. by B.W. Webb) (Proc. Hamburg Symp., August 1983), 41-50. IAHS Publ. No. 141.

Garrel, R.M. and Mackenzie, F.T. (1971). *Evolution of Sedimentary Rocks.* Norton, New York, USA., W.W., Norton, p. 450.

Gibbs, R.J. (1970). "Mechanism controlling world water chemistry." *Science*, 170, pp. 1088-1090.

Hallet B. (1976). "Deposits formed by subglacial precipitation of $CaCO_3$." *Geol Soc Am Bull*, 87:1003-1015.

Hasnain, S.I. and Thayyen, R. (1996). "Sediment transport and solute variation in meltwater of Dokriani glacier (Bamak), Garhwal Himalaya." *Geol Soc India*, 47:731-739.

Kumar, K., Miral, M., Joshi, S., Pant, N., Joshi, V. and Joshi, L.M. (2008). "Solute dynamics of meltwater of Gangotri glacier, Garhwal Himalaya, India." *Environ Geol* (2009) 58:1151-1159.

Lorrain, R.D. and Souchez, R.A. (1972). "Sorption as a factor in the transport of major cations by meltwater from and Alpine glacier." *Quarter Res* 2(2): 253-256.

Puri, VMK. (1994). "Element of glaciology." In *Geological Survey of India.* Foundation course in Glaciology 79.

Raina, V.K. and Srivastava M. (2008). *Glacier Atlas of India,* GSI.

Rainwater, F.H. and Guy, H.P. (1961). Some observation on the hydrochemistry and sedimentation of the Chamberlin glacier area, Alaska. Geological survey of USA professional paper, 414, pp. 1-14

Raiswell, R. (1984). "Chemical models of solute acquisition in glacial meltwater." *J Glaciol*, 30(104): 49-57.

Reynolds, R.C. and Johnson, N.M. (1972). "Chemical weathering in the temperate glacial environment of the Northern Cascade mountains." *Geochim Cosmochim Acta,* 36:537-554.

Sangewar, C. and Bajpai, I.P. (2009). *An Inventory of Himalayan Glaciers,* GSI.

Singh, P., Haritashya, U.K., Kumar N. Sing, Y (2006). "Hydrological Characteristics of Gangotri glacier, Central Himalaya, India", *Journal of Hydrology,* 327, 55-67.

Tranter M., Raiswell, R. (1991). "The composition of the englacial and subglacial components in bulk meltwaters draining the Gornergletscher, Switzerland." *J. Glaciol,* 37(125): 59-66.

Tranter, M., Brown, G.H., Raiswell, R., Sharp, M.J., Gurnell, A.M., (1993). "A conceptual model of solute acquisition by Alpine glacial meltwaters." *J. Glaciol,* 39, 573-581.

Trudegill, S.T. (1986). *Solute Process.* John Wiley & Sons Ltd., Chichester, p. 509.

Index